HPLC Detection
Newer Methods

EDITOR

Gabor Patonay

VCH

Gabor Patonay
Department of Chemistry
Georgia State University
Atlanta, GA 30303-3083

Library of Congress Cataloging-in-Publication Data

HPLC detection : newer methods / Gabor Patonay (editor).
    p. cm.
  Includes bibliographical references and index.
  ISBN 0-89573-327-7 : $95.00
  1. High performance liquid chromatography. I. Patonay, Gabor.
QD79.C454H64 1992
543′.0894—dc20

92-26716
CIP

© 1992 VCH Publishers, Inc.

Printed in the United States of America
ISBN 0-89573-327-7
ISBN 3-527-28219-X

Printing History:
10 9 8 7 6 5 4 3 2 1

Published jointly by

VCH Publishers, Inc.
220 East 23rd Street
New York, New York 10010

VCH Verlagsgesellschaft mbH
P.O. Box 10 11 61
D-6490 Weinheim
Federal Republic of Germany

VCH Publishers (UK) Ltd.
8 Wellington Court
Cambridge CB1 1HZ
United Kingdom

Preface

High-performance liquid chromatography (HPLC) has been one of the most important analytical methods used both in industry and academe. A large number of detection methods are available for detecting the separated solutes, and several of these methods have been used for a long time. Hence, the theory and application of these conventional HPLC detection methods have been discussed in detail in a large number of excellent monographs, review articles, and reports. Some of the more conventional detection methods, such as UV/vis absorption, fluorescence, and conductivity detection, are the detection methods of choice for most analyses. However, with the advent of several new developments in analytical chemistry, new detection methods have become available as HPLC detectors. For example, long-lived luminescence under a variety of experimental conditions may be advantageous as an HPLC detection mechanism that significantly lowers background interference. Also, Fourier transform infrared- (FTIR) and mass-spectrometric techniques have been employed to detect HPLC eluents. From these and other advances in analytical chemistry, a number of novel HPLC detection methods have been developed. In the area of bioanalytical chemistry, several reports have appeared in which long-wave-length fluorescence using the extremely cost-effective semiconductor laser excited fluorescence has been employed to detect biologically important molecules in the HPLC eluent.

The primary purpose in writing this book was to give a rather detailed survey of these new, less conventional detection methods. It is not the goal of this book to provide comprehensive coverage of the detection methods used in HPLC separations; several excellent monographs have already accomplished this. For the sake of brevity, several of the most widely used methods and detectors have been intentionally omitted. Accordingly, the reader will not find detailed discussions about conventional UV HPLC detection or regular fluorescence. However, emphasis has been placed on the more modern or just developing detection methods and their use in HPLC separations. This book is intended to serve the needs of those chromatographers whose work is restricted by the limits of conventional detection methods. It will be of interest to those working in HPLC separations and those who have a more specialized interest in HPLC detection. The book should also find use as a reference text or supplemental material for courses involved with separation analysis techniques.

In Chapter 1 measurement concepts are presented for detection in micro-HPLC separations using lasers. In Chapter 2, the advantages of using long-lived luminescence detection methods are discussed to illustrate its applications with trace concentrations. The utility of chemiluminescence in HPLC detection is presented in Chapter 3. Chapter 4 discusses near-infrared semiconductor laser fluorescence, one of the latest emerging detection methods for ultratrace concentrations. The somewhat more conventional electrochemical detection method is discussed in Chapter 5, however, with a special emphasis on less conventional applications.

Chapter 6 serves as an introduction to the second part of the book, which discusses powerful photothermal detection methods. This detection method has significant utility at extremely low concentrations. The last three chapters focus on detection methods that are supplying a wealth of information about the analyte molecule in the detector. These methods may supply enough information for complete identification of the solute, hence improving the utility of HPLC. Chapter 7 discusses HPLC detection using FTIR spectroscopy. Chapter 8 gives a detailed summary of one of the most powerful detection methods, HPLC mass spectrometry. Finally, Chapter 9 is a peek into the future, indicating how the ultimate power of NMR may be combined with HPLC.

I would like to convey my most sincere appreciation to the authors of the chapters of this book. My appreciation is also extended to Dan Adams for his expert assistance on style and readability, and Dianne Becht for help in the final formatting of the book.

Georgia State University   G. Patonay
Atlanta, Georgia

Contents

5. Electrochemical Detection for Liquid Chromatography   91
J.Wang

6. Photothermal Detectors for High-Performance Liquid Chromatography   111
C. D. Tran

7. HPLC Detection Using Fourier Transform Infrared Spectrometry   127

V. F. Kalasinsky and K. S. Kalasinsky

8. HPLC Detection by Mass Spectrometry   163

K. B. Tomer

List of Contributors

DR. THOMAS J. EDKINS. R. W. Johnson Pharmaceutical Research Institute, Analytical Research and Development, Spring House, PA 19477-0776

DR. DENNIS C. SHELLY. Department of Chemistry, Texas Tech University, Lubbock, TX 79409-1061

DR. DAVID S. HAGE. Department of Chemistry, University of Nebraska, Lincoln, NE 68588-0304

DR. C. GOOIJER. Free University, Department of General and Analytical Chemistry, de Boelelaan 1083, 1081 HV Amsterdam, The Netherlands

DR. M. SHREURS. Free University, Department of General and Analytical Chemistry, de Boelelaan 1083, 1081 HV Amsterdam, The Netherlands

DR. N. H. VELTHORST. Free University, Department of General and Analytical Chemistry, de Boelelaan 1083, 1081 HV Amsterdam, The Netherlands

DR. JOSEPH WANG. Department of Chemistry, New Mexico State University, Las Cruces, NM 88003

DR. CHIEU D. TRAN. Department of Chemistry, Marquette University, Milwaukee, WI 53233

DR. VICTOR F. KALASINSKY. Department of Environmental and Toxicologic Pathology, Armed Forces Institute of Pathology, Washington, DC 20306

DR. KATHRYN S. KALASINSKY. Division of Forensic Toxicology, Office of the Armed Forces Medical Examiner, Armed Forces Institute of Pathology, Washington, DC 20306

DR. KENNETH B. TOMER. Laboratory of Molecular Biophysics, National Institute of Environmental Health Sciences, P.O. Box 12233, Research Triangle Park, NC 27709

DR. KLAUS ALBERT. Institut für Organische Chemie, Auf der Morgenstelle 18, D-7400 Tübingen, F.R.G.

DR. ERNST BAYER. Institut für Organische Chemie, Auf der Morgenstelle 18, D-7400 Tübingen, F.R.G.

CHAPTER

1

Measurement Concepts and Laser-Based Detection in High-Performance Micro Separations

Thomas J. Edkins

The R. W. Johnson Pharmaceutical Research Institute
Analytical Research and Development
Spring House, Pennsylvania 19477-0776

Dennis C. Shelly

Department of Chemistry
Texas Tech University
Lubbock, Texas 79409-1061

1.1. Purpose

High-performance micro separations combine with laser-based detection for maximum analytical performance, which is demanded in such areas as pharmaceuticals, biomedicine, and environmental science. The unique combination of laser radiation and micro separation techniques has as its basis the so-called analytical figures of merit. We devote this chapter to the measurement fundamentals of this group of microanalytical detection schemes. We shall also attempt to outline the current state of commercial development of these techniques.

1.2. Introduction

1.2.1. The Instrument Response Function and Figures of Merit

Probably the most important aspect of laser detection for micro separations is the instrument response function, as expressed by the degree of instrument (detector) response per analyte concentration or mass. Figure 1.1 shows a representative response function. Note that the axes extend over several orders of magnitude and

1

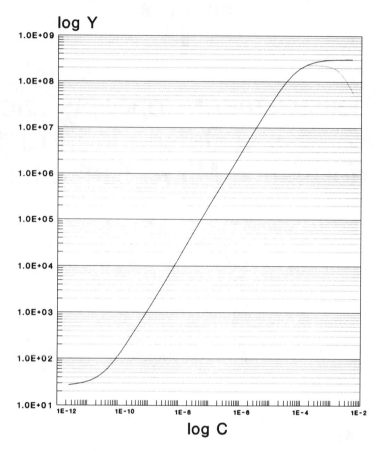

Figure 1.1. Instrument response function as log Y versus log C, where Y is the instrument response and C the concentration of analyte. Analyte mass, M, may be substituted for C. The dotted line indicates a "rollover" effect.

that there are nonlinear portions at both low and high analyte concentration (or mass). The overall instrument response function, which includes these nonlinear regions, is obtained through curve-fitting techniques. The linear portion of the response function[1] is given by

$$Y = AC^r, \tag{1.1}$$

where Y is the instrument response (volts, microamps, etc.), A is a constant that is characteristic of the particular measurement, C is the concentration (or mass), and r is the response index. Taking the logarithm of both sides and simplifying, we have

$$\log Y = \log A + r \log C, \tag{1.2}$$

where the ordinate intercept is log A and the slope is r. Deviation of r from unity implies that the instrument response–concentration (or mass) relationship is skewed positively ($r > 1$) or negatively ($r < 1$). What is ideal about the condition $r = 1$? Simply, that we would expect a 10- or 100-fold increase in response for a 10- or 100-fold increase in analyte concentration (or mass). This direct proportionality implies perfect correlation between input and output (i.e., chemical information and analytical signal.).[2] Overall, the instrument response function is sigmoidal, with a positive deviation from linearity at low concentration (mass) and negative deviation at high concentration (mass). These deviations are due to both measurement and instrument characteristics, such as analyte photophysics and the arrangement of electronic components within the instrument. These aspects will be discussed in the following. While the entire measurable range would include these deviations, the analytically useful portion is restricted to the linear region.

The range in concentration (or mass) over which this response function is linear is known as the *linear dynamic range*. It is properly defined as the range in analyte concentration (or mass) for which the response index is between 0.98 and 1.02. In other words, we restrict the direct proportionality between analyte concentration (or mass) and instrument response to ±2% deviation from 1 for an "operational" convenience. The linear dynamic range is properly determined using statistical routines whereby limits can be placed on the slope (r) and the largest abscissa range that satisfies this criterion is found.

The instrument response time is another important feature of laser-based detection. The time required for the output signal to increase or decrease to $1/e$ of the maximum response value is the response time. With no input signal the output will lag slightly behind an input signal pulse ("on"), as shown in Figure 1.2. Similarly, the output will lag temporally behind an "off" transition at the input. While this is classically attributed only to transducers, this response characteristic should be ascribed to transducers plus their associated electronics. The nature of this "signal processing" by detectors and their electronics can be linear or exponential with respect to time. Typically, analog electronics display exponential signal–time relationships, while digital electronics show a linear dependency.[3]

The slope of the calibration curve, a selected, small portion of the response function, is defined as the *calibration sensitivity*. If the sensitivity is compensated by the measurement precision, then it is referred to as the *analytical sensitivity*.[4] Presumably, if one had constructed a response–concentration calibration, then the slope of this plot would be called the *concentration calibration sensitivity;* likewise, for response–mass calibration, the slope would be the mass calibration sensitivity. The important concept here is the degree of proportionality between ordinate and abscissa for both these plots. Simply, the degree of change in these quantities is the sensitivity. Obviously, the greater the slope, the greater will be the change in ordinate value per abscissa increment. The utility of calibration and analytical sensitivities in analytical separations stems from the fact that no two analytes will have identical sensitivities, owing to differences in detector response and separation efficiency and resolution. The bottom line is that each analyte should be separately calibrated.

As one approaches the lowest response levels of the instrument, one encounters a nonlinear portion, which determines the limit of detection. The limit of detection is

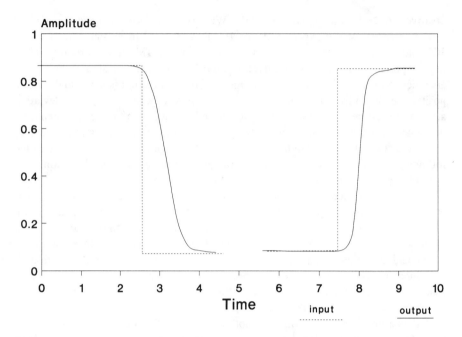

Figure 1.2. Temporal response of a detection system to negative and positive transients. ···
input signal; ——, output signal.

a parameter that is defined as the minimum distinguishable signal that is two or
three times higher than the standard deviation of the baseline noise.[4,5] There is some
disagreement[5] as to whether there must be linearity as one approaches the detection
limit; hence, the detection limit might be roughly equal to the minimum concentration (or mass) in the linear dynamic range computation. For safety's sake, one
would choose three standard deviations as the limit of detection, and this roughly
corresponds to this point in the linear dynamic range. In any case, it is generally
agreed that detector or baseline noise determines the limit of detection. Detector
noise is largely independent of the type of source used. Laser-based detectors often
display higher signal-to-noise levels, thus enabling lower detection limits to be
realized, mostly because of the greater signal that is generated rather than by any
reduction in noise.

These parameters—instrument response function, linear dynamic range, response time, calibration and analytical sensitivity, and limit of detection—are commonly referred to as *analytical figures of merit*. It is well known that lasers are ideal
spectral sources for detectors and that micro separations are ideal analytical separation techniques, each in terms of its performance criteria. Yeung has eloquently and
completely presented this case.[6,7] Logically, there should be unique aspects to the
analytical figures of merit for laser-based micro separation techniques. Figure 1.3
shows the attributes of both the source (laser) and separation technique that influence the analytical figures of merit. More important, however, Figure 1.3 shows the

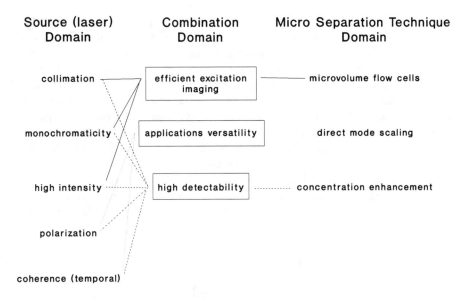

Figure 1.3.   Comparison of attributes for the laser source, micro separation technique, and the combination.

combinations of attributes that influence detection performance with laser-based micro separation detectors. Especially unique to micro separations is the need for microvolume flow cells, an attribute that combines very well with the collimation, monochromaticity, and high intensity of the laser to provide for very efficient excitation imaging. Finally, applications versatility results when all the laser source attributes are combined with the fact that micro separations are a miniaturized version of the conventional (larger) equivalent. Indeed, one simply has to use the same sorbent and mobile phase in the miniaturized separation as was required in the large-scale version.[8] Selectivity and capacity ratios are mostly equivalent, allowing the same type of separation to be accomplished but on a much smaller (micro-analytical) scale. High detectability is an attribute of the combination that is due to the monochromaticity, collimation, high intensity, polarization, and coherence of the laser source, combined with the concentration enhancement achievable with microscale analytical separations. Specific examples of these attributes and their impact on the analytical figures of merit will be discussed in the following.

1.2.2. Concentration and Mass Sensitivity

There are other analytical figures of merit related to sensitivity that are intended to quantify the mass and concentration response of detectors. Scott presents the terms *detector sensitivity* and *mass sensitivity* of chromatographic detectors, referring not to the slopes of the appropriate calibration plots but to the limits of detection to which the instrument responds to concentration and mass.[9] This argument is based on the fact that the signal-to-noise ratio is highest at the apex of a chromatographic peak, where presumably the concentration or mass flux is also highest. Because of the very small elution volumes, connecting tubing and flow cells of micro separation techniques, there is "increased mass sensitivity of concentration-sensitive detectors" in these techniques.[10] There need not be confusion, when one realizes that some detectors respond more to number density of analyte and others to number or mass flux of analyte. Since concentration and mass are interconvertible through the volume, mass sensitivity depends on concentration sensitivity to a large degree. With micro separations one would like to preserve, as much as possible, the concentration response of detection because solute mass becomes vanishingly small approaching the detection limit.

A fundamental constraint to all concentration-sensitive detection strategies is the path length, as determined by Beer's law. The illuminated volume (as determined by

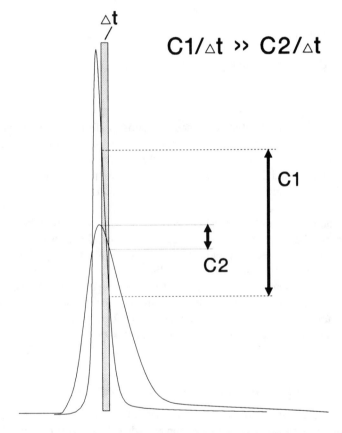

Figure 1.4. Illustration of enhanced mass sensitivity in high-efficiency separations using concentration-sensitive detection. The more efficient chromatogram shows a larger concentration difference, C_1, hence a larger mass flux of analyte during the sampling interval, Δt.

path length and capillary diameter) must be very small for micro separations; thus, concentration-sensitive detection will likely be degraded relative to mass-sensitive detection. The only solution is to make full use of the available cell volume.

The influence of separation efficiency on detection is illustrated in Figure 1.4. Two concentration profiles (peaks) represent identical injected solute mass but different separation efficiency. If the sampling time is the same in both cases, it should be obvious that the higher-efficiency peak presents a higher concentration and mass flux of analyte to the detector, compared to the lower-efficiency peak. For the detector that responds to concentration, there will be a greater mass flux through this detector with the faster, narrower peak than the slower, broader one. The output will display the same response to concentration, but now, increased response to mass. Therefore, the performance of the micro separation clearly enhances detection performance with respect to both concentration and mass response (sensitivity).

In laser-based detectors the key issue with respect to concentration and mass sensitivity is the type of measurement that is made and not the type of source that is used. The nature of the transducer (photomultiplier tube, photodiode, etc.), as well as analyte photophysics, are principally involved. Some measurements such as absorption and fluorescence are intrinsically concentration based. Others, such as photoionization and mass spectroscopy, are more clearly mass sensitive. Ion currents and ion count rates are nearly exclusively linked to the flux of ions or ion-producing species. Laser-based detectors for micro separations display enhanced excitation efficiency in nearly all measurement scenarios because of the unique compatibility of attributes illustrated in Figure 1.3. This imparts greater performance to both concentration- and mass-sensitive detection compared to non-laser-based conventional-scale separation strategies.

1.2.3. Optical Imaging and Detection Strategies

By far the most commonly used technique for imaging a laser onto separated solutes for the purpose of analytical detection is to focus the beam with lenses onto an optically transparent capillary, through which the solute and mobile phase flow. Yeung has discussed the fundamentals of imaging lasers onto such miniature flow cells for chromatographic detection.[6] Photolysis, photothermal effects, and optical defocusing are some of the detrimental influences of the optical configuration and fluidics of flow cells. The importance of flow cell design and operation in laser-based micro separation detection has been considered by both McGuffin and Zare[11] and Poppe.[12] Some of the important issues discussed were geometry, flow profiles, and materials considerations.

With respect to detection, there are many different imaging strategies, owing to the variety of measurement types and optical configurations. The use of high-numerical-aperture lenses and full transducer illumination strategies, both serving to maximize the signal-to-noise ratio of the optoelectronic information, are the general themes. Flicker noise is a common noise component of laser systems. Most detection systems for laser-based micro separations employ some type of signal-to-noise enhancement approach, such as modulation, lockin amplification, or dynamic referencing. Digital signal processing techniques must be used appropriately so that solute or chemical information is not lost through attenuation of the higher-frequency components from the separation.

1.2.4. Direct and Indirect Detection

It is possible to detect the presence of analyte directly or indirectly. Figure 1.5 shows a comparison of absorption and emission measurements made with both direct and indirect detection strategies. Indirect detection can be regarded as displacement detection because the analyte displaces a probe species in the mobile phase, producing diminished signal intensity, because the probe rather than analyte is directly measured.[13]

In Figure 1.5A we see the typical direct absorbance measurement as the very small difference between two large optical signals. In the absence of analyte, there is no difference recorded between I_0 and I. When a small amount of analyte is present, I decreases to an almost imperceptible degree, even as the log ratio of I to I_0 is computed. Absorption can be measured indirectly, however, using an indirect

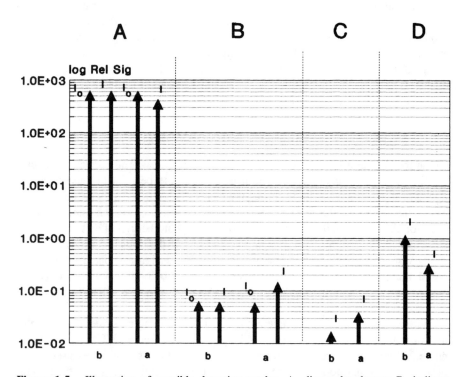

Figure 1.5. Illustration of possible detection modes. A. direct absorbance; B. indirect photometry; C. thermal lens, polarimetry, and fluorescence; D. indirect fluorescence. Key: I_0, excitation light signal; I, detected light signal; b, before or without analyte; a, after or with analyte.

photometric technique. This is shown in Figure 1.5B. Here, the presence of probe drastically diminishes the light intensity for both I_0 and I. When analyte displaces probe, the light intensity increases, resulting in enhanced I, as shown. Better discrimination between background and analyte is recorded because probe absorption is characterized by higher calibration sensitivity than that for the analyte. Improved sensitivity and limit of detection result. Because these are absorption measurements, two readings (I_0 and I) are required for each measurement (before and after analyte).

Emission measurements are also possible by direct and indirect detection. Figure 1.5C shows the scenario for direct fluorescence. Here, a single measurement is required in both the before and after analyte situations. Note the magnitude of the signals relative to those for absorption. Direct analyte detection is accomplished with a good signal-to-noise ratio because signal intensity measurements are easily and efficiently made. Indirect fluorometric detection is accomplished by a measurement of probe fluorescence. This is shown in Figure 1.5D. Now, as probe is displaced by analyte, less signal intensity is recorded. Obviously, nonfluorescent analytes are detected in this manner. As in indirect photometric detection, the probe calibration sensitivity is greater than that for the analyte; so a better measurement results. Both nonabsorbing and nonfluorescing analyte species are amenable to indirect detection strategies. Additional details of these methods will be discussed.

1.2.5. Varieties of Micro Separations

With the advent of many forms of capillary or micro separation techniques, there has been an attempt to unify the various formats and theoretical nuances into a broad, all-inclusive concept of analytical separations. This has been attempted recently by Giddings,[14] Novotny,[15] and Yang.[16] Table 1.1 shows a comparison of the column dimensions and operational features of the major types of micro separation techniques to which laser-based detectors have been applied. Their most common feature is the use of glass or fused silica capillaries with a circular cross

Table 1.1. COMPARISON OF CAPILLARY SEPARATION TECHNIQUES

Technique	Column i.e. (μm)	Length (m)	Packing	Flow rate (nL/min)	Peak vol. (nL)
HPLC	4600	0.25	yes	2×10^6	1×10^6
μ-column	250	0.5	yes	2000	1000
Pkd. capl.	75	1.0	yes	300	100
OTLC	5	1.5	no	10	10
CZE	80	0.75	no	10	2
HPCE	100	0.25	yes	30	100
CITP	100	0.25	no	*	100

Abbreviations: HPLC (high-performance liquid chromatography); μ-column (microcolumn); Pkd. capl. (packed capillary); OTLC (open tubular liquid chromatography); CZE (capillary zone electrophoresis); HPCE (high-performance capillary electrophoresis); CITP (capillary isotachophoresis). *Flow rate not applicable.

section. As shown, inner diameters (i.d.) range from less than 10 to slightly more than 300 μm and column lengths range from several centimeters to several meters.

1.3. Laser-Based High-Performance Micro Separation Measurements

1.3.1. Review of Trends

Laser-based detection for liquid chromatography is a "maturing" research technique, as evidenced by the abundance of literature over the past 15 years.[17–19] Yet lasers have been incorporated into very few commercial HPLC detectors. Many recent instrument developments are specifically aimed at bridging the gap between research and commercialization.

One current trend that may be a driving force for laser-based detector development is the miniaturization of separation schemes. While both μ-LC and capillary electrophoresis (CE) instrumentation can often achieve efficiencies (N) $> 1 \times 10^5$, there is no detector that provides adequate detectability for most analytes. Many instruments come equipped with UV detectors originally designed for conventional HPLC (3–4.6 mm i.d. columns). Due to path length constraints imposed by Beer's law, detectability is often 10–50× higher for a micro separation. Employing either an axial-illuminatcd[20] or a Z-shaped flow cell[21] has been suggested to overcome this limitation.

Another trend is that many investigators are looking to laser-based detection schemes to find a truly "universal" detector for μ-LC and CE. Many novel designs have appeared recently, such as magnetooptical rotation[22] and concentration-gradient detection.[23] Perhaps the most promising universal schemes for micro separations are found among the indirect detection modes, which have been studied by several investigators.[24–26]

A third trend is that lasers are slowly being introduced into commercial instruments for both HPLC and micro separations. One of the earliest examples is the low-angle laser light scattering detector (LALS), which is useful for characterizing macromolecules separated via gel-permeation chromatography.[27] A second example is the diode-laser-based polarimeter, for detection of optically active species.[28] A laser-induced fluorescence (LIF) detector for CE has been introduced, for identifying amino acids and peptides using fluorescent dye labeling.[29]

Perhaps the most impressive application to date of a laser in a micro separation is in support of the human genome initiative,[30,31] an international, multidisciplinary, 15-year project intended to provide a completely sequenced genetic map of a human sperm (or egg) cell. Variations of CE, coupled with LIF detection of derivatized bases, have been proposed for DNA sequencing. Dovichi has suggested using up to 50 capillaries in parallel[31]; another approach is to use pulsed-field capillary gel electrophoresis.[31] This particular laser-based micro separation technique may influence future research in molecular biology.

1.3.2. Approaches to Selective Detection Using Lasers

Selective laser-based detection schemes for micro separations encompass four branches of optical spectroscopy: absorbance, fluorescence, optical activity, and scattering. The utility of lasers for each of these techniques can vary widely, based on the fundamental spectroscopic properties of each measurement. Refractive index (RI) is defined as a universal detection technique and will be covered in a later section.

1.3.2.1. Absorbance

Because a highly monochromatic laser lacks the tunability of a D_2 or W_2 source, laser-based absorption is a selective detection mode. Lasers can theoretically improve detectability by either extending the shot-noise limit at low solute concentrations, or by increasing the dynamic range at high solute concentrations.[32] In reality, scattering within the flow cell and/or source flicker noise often limits any potential sensitivity gains. Nor surprisingly, very few examples of laser-based absorbance are found in the literature.

Perhaps a more promising approach is to measure absorption by sensing associated processes, such as photoacoustic or thermal lens detection.[33,34] These techniques have been defined as "indirect" absorbance measurements by Morris.[35] Because the signal itself is measured directly (as in fluorescence), it is less subject to stray light effects and scattering losses than direct absorbance detection. Indirect absorbance is not associated with the concept of indirect (photometric) detection. The former applies to sensing absorption-related phenomena, such as thermal lensing, while the latter refers to direct measurement of probe displacement by analyte (an indirect measure of analyte presence). Indirect photometric detection and indirect absorbance are illustrated in Figure 1.5B and 1.5C, respectively.

The photoacoustic effect measures the pressure change of a solute as an acoustic wave. Although there was some interest in this technique for conventional HPLC in the early 1980s,[34,36] it has not been widely used because of problems with heat dissipation, convection in liquids, and difficulties with the liquid–detector interface.

Use of the thermal lens detector is still being pursued today. The formation of the thermal lens is almost identical in appearance to a diverging lens in geometrical optics, as shown in Figure 1.6. A thermal gradient occurs, most pronounced at the beam center, during the relaxation process following absorption. If the detector flow cell is placed one confocal length from the beam waist, the resulting signal can be quantified as:

$$\frac{\Delta I_{bc}}{I_{bc}} = -2.303 \left[P \left(\frac{dn}{dT} \right) / \lambda k \right] A, \tag{1.3}$$

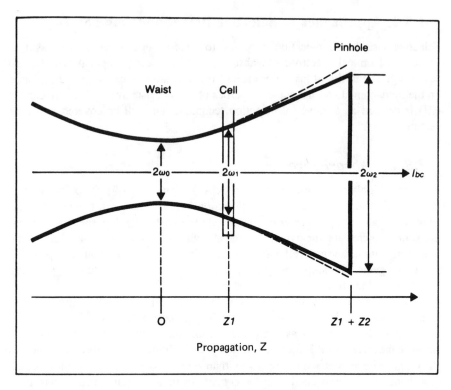

Figure 1.6. Optical system for detecting the thermal lens effect. A sample cell is placed a distance Z_1 from the waist. Z_2 indicates the distance between the cell and a pinhole, which samples the intensity at the beam center, I_{bc}. The spot size of the beam at the waist, sample cell, and pinhole are w_0, w_1, and w_2, respectively. (Reprinted with permission from Ref. 33. Copyright 1980 American Chemical Society.)

where I_{bc} is the beam intensity at the center, P the laser power, dn/dt the change in RI with respect to temperature, k the coefficient of thermal conductivity of the bulk solvent, and A the sample absorbance. Equation (1.3) can be reduced to:

$$\Delta I_{bc}/I_{bc} = 2.303EA, \tag{1.4}$$

where E is called the enhancement factor, which is the collection of terms within the brackets in Eq. (1.3).

Thermal lens instruments often employ both a pump and probe laser, typically Ar^+ and He-Ne sources, respectively. Dovichi has recently suggested that for a crossed-beam thermal lens detector, the signal is inversely proportional to the pump laser beam size.[37] Therefore, this design has high utility for capillary separations. Applications of thermal lensing for μ-LC[38,39] and CE[40] have been demonstrated. Typical detection limits were 750 attomoles injected, representing a concentration of $1.3 \times 10^{-8} M$, for 4-4-Dimethyl Amino Azo Benzene-4′ Sulfonyl-chloride (DABSYL) derivatized amino acids. The theoretical concentration limit at the detector was estimated to be $4 \times 10^{-10} M$, or 50 molecules in the illuminated volume of the cell!

An often overlooked aspect of "direct" absorbance measurements is that at low solute concentrations (0.004–0.100 abs.), the absorption as seen by the detector is actually a small difference in two large signals. This is illustrated in Figure 1.5A. When a molecule absorbs a small amount of the incident light (I_0), a decrease in photon flux is registered by the transducer (Figure 1.5A, *a*). A direct absorption measurement actually displaces light intensity. In indirect absorbance measurements, the situation is reversed. Light intensity for a blank measurement approaches zero, as shown in Figure 1.5C, *b*. In Figure 1.5C, *a* represents the analytical signal during thermal lensing (or the photoacoustic effect). The light intensity at the detector is directly proportional to the solute concentration.

Figure 1.5C represents an important concept in estimating the utility of lasers for detection techniques: It is intuitively obvious that it is easier to measure a small signal against a low background, rather than a small signal against a high background. This is a primary reason that lasers have been so readily adaptable to low-light-level techniques such as thermal lensing, fluorescence, and polarimetry.

1.3.2.2. Fluorescence

Laser-induced fluorescence (LIF) has been the most frequently used and successful of all laser-based techniques. Lasers are ideal for fluorescence detection due to the well-known direct proportionality between the lasting source and the resulting analytical signal:

$$I_f = I_0 (2.303) Ebc\Phi_F k, \tag{1.5}$$

where I_f is the intensity of the fluorescent signal, I_0 the intensity of the incident signal, Φ_F the solute fluorescence quantum efficiency, k the instrument fluorescence collection efficiency, and the other terms are as in Beer's law. Figure 1.5C shows direct fluorescence detection before (*b*) and after (*a*) analyte has entered the flow cell.

The use of lasers for fluorescence excitation has a profound impact on detectability. From Eq. (1.5) we would expect much higher I_f in laser-induced fluorescence than with conventional sources. However, this improvement in detectability must be tempered against the relative flicker noise characteristics of laser and conventional sources such as D_2 and Xe lamps. Laser-based fluorescence detection is often background limited by source flicker noise[32] or scattering processes within the interrogation region.[32,41] Other laser-based fluorescence techniques, such as multiphoton ionization[42] and two-photon fluorescence,[43] provide unique selectivity advantages.

Much effort has been devoted to reducing the scattering through improved flow cell designs. McGuffin has rigorously discussed several options, such as the flowing droplet cell, the ensheathed flow cell cuvette, and longitudinal and transverse excitation imaging.[41] Variations of these designs continue today.

The use of LIF in micro separations has a rich history. Applications of the technique for open tubular LC,[44] CE,[45,46] and packed microcolumn LC[47,48] abound in the literature. Impressive, and even spectacular, detection limits have been reported.[49-54] Dovichi et al. have reached single-molecule detection limits for rhodamine-6G, using flow cytometry.[51] In this work the linear dynamic range was more than 5 orders of magnitude, with a response index of 1.000 ± 0.009 and a concentration detection limit of $8.9 \times 10^{-14}\ M$.

Many studies have indicated the improvement in mass sensitivity as a result of miniaturization,[52-54] but the effect on concentration sensitivity is not as straightforward. A recent review states that concentration sensitivity suffers when employing a miniaturized flow cell because of shortened optical path length, as compared to conventional HPLC.[32]

We have recently compared the system performance of LIF detection for μ-LC to a commercial fluorometer for HPLC.[53] Both minimum detectable quantities (MDQs) and minimum detectable concentrations (MDCs) were compared between the two instruments, for both pyrene and a preliminary measurement of derivatized amino acids. The μ-LC/LIF system gave markedly superior MDQs for both solutes. The HPLC/fluorometer gave slightly lower MDCs for pyrene, but much higher ones for the derivatized amino acids.[50]

The concentration sensitivity improved for the amino acids because the lasing wavelength (325 nm) was very close to the excitation maximum, while for pyrene it was not. Results from another recent study seem to support these data.[55] The improvement in MDC in the μ-LC/LIF experiment for the one solute may be attributed to the higher excitation energy, resulting in increased fluorescence [I_o and I_f, respectively, in Eq. (1.5)]. This implies that the shorter path length in a micro separation may not be the ultimate limitation to the concentration sensitivity. Rather, adventitious fluorescence from transition metal impurities in the fused silica, Raman scattering from the mobile phase, and/or dark current from the photon transducer contribute more extensively than the short path length.

1.3.2.3. Optical Activity Measurements

Measurement of optical activity is often used for enantiomeric purity determinations of chiral species, which by definition have either a center or plane of asymmetry. The degree of optical rotation can vary with wavelength, optical path length, temperature, concentration, and magnetic field strength.[56]

Often only one enantiomer of an enantiomeric pair is biologically active. A striking example can be seen in the development of the drug thalidomide, which was marketed as a racemate in Western Europe in the early 1960s. Thalidomide has been shown to cause phocomelia in infants, which is the attachment of the hands or feet directly to the shoulder or hip.[57] Strong, but inconclusive, evidence suggests that this effect occurs in only one enantiomer.[58] This tragedy may have been avoided if a stereospecific synthesis had been available at the time of development.

Lasers have been incorporated into two optical rotation methods to date: polarimetry and circular dichroism (CD). A polarimeter measures the direction of rotation of plane-polarized light caused by an optically active substance. The degree of rotation can be measured by:

$$\alpha = [\alpha]\ 100bc, \tag{1.6}$$

where c is the solute concentration in grams per 100 ml of solution, b the optical path length in dm, α the observed rotation in degrees, and $[\alpha]$ the specific rotation.

Circular dichroism measurements provide both absorbance and optical rotation information simultaneously.[59] A spectrophotometer is used to measure the differential absorption of circularly polarized light:

$$\Delta A = cb\ \Delta E, \tag{1.7}$$

where b is the path length and $\Delta E = (E_L - E_R)$, the differences in rotation between left and right circularly polarized light.

The concept of laser-based polarimetry for HPLC was initiated by Yeung and co-workers, ostensibly because attempts at interfacing commercial polarimeters with HPLC units gave poor detectability.[60] Recently, mass detection limits of 5 ng for penicillin G has been reported with this detector.[61] A schematic of their design is shown in Figure 1.7.

The rationale for using lasers in a polarimetric detector is several-fold. Use of a laser improves the polarization extinction ratio (ratio of polarized light to unpolarized light) of the polarizer/analyzer prism combination ca. 10,000×.[60] Second, crossing the polarizer and analyzer prisms at a 90° angle provides a low light level for a blank measurement. This is an ideal optical condition for laser-based detection (Figure 1.5C). Third, use of the laser permits imaging into small flow cell volumes (< 1 µl); hence, 11 ng detection limits and linearity of approximately three orders of magnitude have been achieved for fructose with microbore LC.[62]

Laser-based CD detection is a more selective optical activity measurement than laser-based polarimetry. Because it is inherently an absorption technique, it is necessary to employ a solute that absorbs at or near the lasing wavelength. A CD measurement is actually a small difference between two large signals, similar to the diagram shown in Figure 1.5A.

The principal reason for designing a laser-based CD instrument is to improve the mass and differential detectability over interfacing a conventional CD apparatus to HPLC.[63] Polarization modulation at high frequencies (> 100 kHz) is necessary to overcome the high flicker noise present in most lasers. Mass detectability of 5.6 ng has been obtained for microbore LC[63]; detection limits as low as 170 pg have been obtained for HPLC, when CD is combined with laser-induced fluorescence.[64]

1.3.2.4. Scattering

Light scattering measurements can provide physical characterization of chemical species, such as diffusion coefficient and particle size.[65] These methods are conveniently classified by the frequency shift of the resultant scattered light, relative to the incident frequency: (1) elastic (or Rayleigh) scattering, in which there is no net

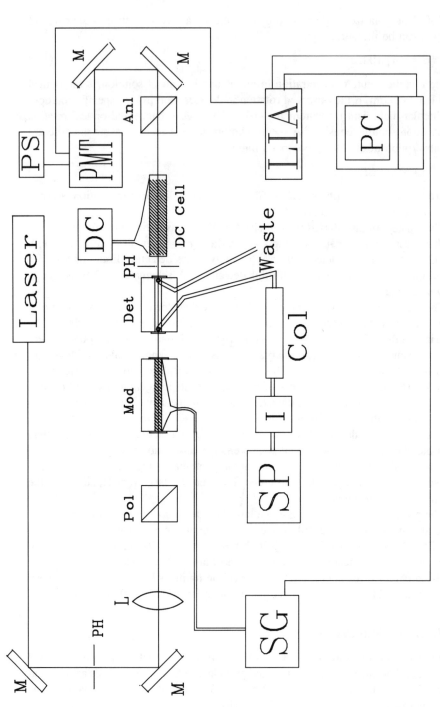

Figure 1.7. Schematic for laser-based polarimetric detector. M, mirror; PH, aperture; L, focusing lens; POL, polarizer; MOD, modulation cell; DET, detection cell; DC, dc power supply (standard Faraday signal); ANL, analyzer; PMT, photomultiplier tube; PS, high-voltage power supply; LIA, lockin amplifier; PC, personal computer; SG, signal generator; SP, syringe pump; I, injection system; COL, HPLC column. (Figure courtesy of Patrick D. Rice.)

16

change in frequency; (2) quasielastic scattering (QELS), which is the measurement of a small frequency shift induced by Brownian motion, and (3) inelastic (or Raman) scattering, which produces a significant frequency shift.

The intensity of elastic scattered light (I_s) is dependent on many physical factors:

$$I_s = 8\pi^4 \alpha r^2/\lambda^4 r^2 (1 + \cos^2 \theta)I_o, \qquad (1.8)$$

where α is the particle polarizability, θ the angle between the incident and scattered light, λ the wavelength, and r the distance between the scattering center and the detector. The cosine dependence of θ and I_s implies that only low scattering angles ($\leq 10°$) will result in significant intensity. Additional requirements are that the laser beam size must be larger than the solute particle size (d_p), and d_p should be between 1.1 and 1.5λ.

Both the low-angle light scattering (LALS) detector and the evaporative light scattering detector (ELSD) are elastic measurements. When used in series with a concentration-sensitive detector, LALS can determine molecular weight measurements of polymers through the Zimm-plot method.[27] By employing the sheath flow cuvette, a μ-LC system has been constructed that detects particles as small as 45 μm.[37] Because the ELSD is considered a universal detector, it will be covered in another section.

A QELS detector provides molecular weight and particle size information of macromolecules in dilute solution.[65] Because it often takes 5–15 min to obtain good spectra of many solutes, it is only used as an off-line detector for gel-permeation chromatography.

The major limitation to applying Raman detection for separations is that the eluent has a Raman signal about 10^3 times stronger than any solute.[66] To improve detectability, signal-enhancing techniques such as surface-enhanced Raman [67] and resonance Raman[68] spectroscopies have been used. Derivatization is often employed to match the solute excitation maximum to the laser Raman line. Many studies using Raman detection for chromatography show the modification of instruments that were originally designed for static measurements. Because a Raman spectrum can provide complete vibrational information, it is potentially a powerful qualitative tool for micro separations.

1.3.3. Approaches to Universal Detection Using Lasers

To date, no HPLC detector has approached the universality of detection that the flame ionization detector (FID) provides for gas chromatography. Some approaches in universal HPLC detection are the seminal work on low-wavelength UV measurements by Berry[69,70] and others.[71,72] Limitations to these approaches include restrictions in the choice of solvents, buffers, and detectors. Often unwieldy and exotic experimental apparatus are employed to purify both solvents and solutes.[70]

The refractive index detector is often described as having a universal response. However, it cannot be utilized for gradient elution, nor is the concentration sensitivity adequate for many solutes.

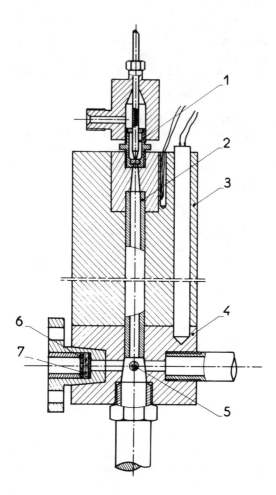

Figure 1.8. Evaporative light scattering detector block. 1, Nebulizer; 2, drift tube; 3, heated copper block; 4, light scattering cell; 5, glass rod; 6, glass window; 7, diaphragms. (Reprinted with permission from Ref. 73.)

1.3.3.1. Evaporative Light Scattering and Refractive Index

Lasers can be used for universal detection when measuring the bulk properties of molecules, such as the evaporative laser light scattering detector (ELSD). The basic design is shown in Figure 1.8. The column effluent is transported from a nebulizer into a short drift tube via an inert gas, which enables a phase separation of the involatile solute from the volatile mobile phase. The resulting analytical signal is proportional to both a power of the solute concentration and its droplet diameter.[73]

The primary advantage of the ELSD over RI detection is that gradient elution is readily achievable. Because the response of this detector seems to be relatively independent of solute type, quantification of each component in a sample mixture is simplified.[73] This detector has been successfully interfaced with μ-LC.[74]

The ELSD does have significant limitations as a universal detector for micro separations. It would probably not be very useful for CE, where involatile buffers and micelles are routinely used. Use of the RI detector imposes even stricter limitations on solvents and solutes.

These restrictions fueled the investigation of alternative approaches. Perhaps the most promising is indirect detection, which was first utilized for detecting ions (e.g., Na^+) that are transparent in the UV.[24] An excellent review of this approach has been given by Yeung.[13]

The utility of any indirect mode can be evaluated by:

$$C_{lim} = C_m/(DR \times TR), \qquad (1.9)$$

where C_{lim} is the concentration limit of detection, C_m the concentration of the mobile phase component that is "affected" by the displacement, DR (dynamic reserve) the ratio of background signal to background noise, and TR (transfer ratio) the number of mobile phase molecules displaced by an analyte.

Inspection of Eq. (1.9) reveals that an ideal indirect detection scheme will have a large DR and TR, low C_{lim}, and relatively low C_m.[13] A pictorial description of indirect detection is shown in Figure 1.9. RI in its "normal" mode of operation is an indirect detection technique, because the RI of the eluent is being displaced by a

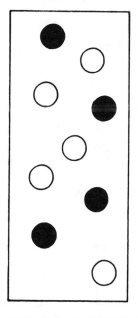 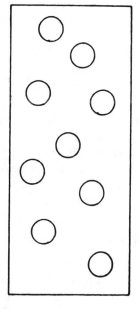

Figure 1.9.   General scheme for indirect detection. Dark circles represent displacement of analyte; lighter circles represent solvent signal. (Reprinted with permission from Ref. 13. Copyright 1989 American Chemical Society.)

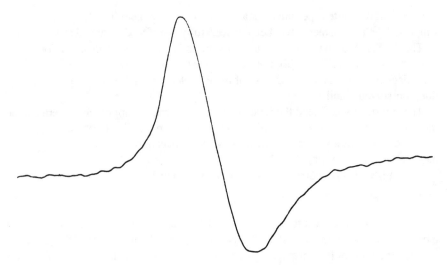

Figure 1.10. Universal concentration gradient signal for low-intensity CW excitation beam. 10 nl injection of $8.1 \times 10^{-5} M$ methylene blue in water, 0.5 mW He-Ne laser beam. (Reprinted with permission from Ref. 23. Copyright 1988 American Chemical Society.)

solute within the interrogation region during detection. Because C_m is large and TR is 1, a large DR is necessary to ensure adequate detectability. RI can be useful as an indirect mode because a DR of 10^7-10^8 is achievable.

An intriguing RI detector design has been introduced by Pawliszyn, which detects the concentration gradient of a solute and not the concentration itself.[23] The sensitivity of this method can be calculated by:

$$\left(\frac{dc}{dx} \right)_{max} = \frac{M}{F} \frac{e^{1/2}N}{t_r^2} \left(\frac{dx}{dt} \right) (2\pi)^{1/2}, \tag{1.10}$$

where $(dc/dx)_{max}$ is the concentration gradient at the maximum, M the total mass injected, N the chromatographic efficiency, F the flow rate, t_r the inflection point, and (dx/dt) the linear velocity. Because detectability is proportional to column efficiency, this detector is ideal for micro separations. A typical chromatogram is shown in Figure 1.10, which shows the peak shape to be identical to the first derivative of a Gaussian curve.

1.3.3.2. Indirect Detection Options

Other indirect modes, such as indirect absorption, can function via a volume displacement. It provides even poorer analytical utility than RI due to a much lower DR (ca. 10^4). Yeung has recently shown that for nonionic indirect absorption measurements, C_{lim} in Eq. (1.9) is ca. $10^{-3} M$.[13] Such a concentration would likely overload most HPLC columns.

Indirect photometric detection has much greater analytical utility because the transfer ratio is dictated by a charge displacement. Indirect fluorescence detection in capillary electrophoresis is perhaps the most ideal example of this strategy. Because capillary zone electrophoresis (CZE) separates solutes by electrophoretic mobility differences, one can use dilute concentrations of ionic fluorophores in the run buffer.[75]

Indirect laser-induced fluorescence is ideal for CE, where very short path lengths (< 50 μm) often limit detection. Concentration limits of detection (LODs) of 10^{-7} M for nucleotides have been obtained.[75] This technique has also been shown to be useful for amino acids and inorganic ions.[76]

The utility of CE can be extended to neutral species by utilizing micellar electrokinetic capillary chromatography (MECC). By employing an ionic surfactant, such as sodium dodecyl sulfate (SDS) in the buffer solution, separation occurs via a partitioning between the micelle and the bulk aqueous phase.[77] Recent work with hydro-organic modifiers in MECC has resulted in further gains in selectivity.[78,79]

Indirect fluorescence of ions in MECC is useful when the fluorophore partitions between the micelle and the bulk phase. Adding organic solvent (either analyte or modifier) will naturally alter the partitioning. As is typical in most indirect modes (e.g., indirect fluorescence), the large background signal causes a large loss in detectability as compared to direct detection.[13,76] The hydro-organic modifier, the probe species, and the micelle all affect the indirect fluorescence measurement; their interrelationships are currently being studied.[76]

While MECC is perhaps the ideal "universal" separation scheme in CE, there is a detection problem for many analytes. The utility of indirect fluorescence is unproved for both neutral molecules and those that do not easily ionize (e.g., sugars). A more universal detection scheme must be developed to extend the utility of micro separations.

The most universal laser-based detection technique may be an indirect detection mode that operates on a volume displacement basis. Additionally, the technique should be easily scaled down to the column i.d.'s (25–250 μm) and flow cell volumes (10–100 nl) typical of a micro separation. It seems that only indirect polarimetry and indirect thermal lens detection can currently meet these requirements.

Indirect polarimetry for microbore LC was investigated in the mid-1980s.[62,80] By employing an optically active eluent in the mobile phase [e.g., R-(+)-limonene], the resulting detection scheme becomes universal for most optically inactive species.[80] The direct detection scheme is selective, of course, for compounds with optical activity.

Surprisingly, in contrast to the previously described indirect modes, indirect polarimetric detection is a low-light-level technique. For both indirect and direct polarimetry the analyzer prism is simply rotated in either mode to null out any background. Detectability in each mode is roughly equivalent.[62]

Indirect polarimetry also possesses a large dynamic reserve. By substituting a specific rotation of 100° and a rotational sensitivity of 10^{-5} deg into Eq. (1.6), a

DR of 10^7 is obtained. This is far larger than the DR for indirect photometry, fluorescence, and absorbance.[13]

The difficulty in implementing indirect polarimetry for micro separations may be in maintaining a high polarization extinction ratio within the flow cell. Approaches such as using fiber optics for increased optical throughput, or extending the path length with an axial-illumination or Z-shaped flow cell, may not work because distortion of the beam wave front will cause depolarization.[81]

In contrast, direct thermal lens detection has been successfully implemented for both CE[37] and μ-LC[38] for many years. It is similar to fluorescence in that detectability is directly proportional to the incident source power. Because the intersection of a pump and probe laser produces very small volumes (ca. 0.1–10 nl), the technique is ideal for imaging into small capillaries.

The concept of indirect thermal lens detection has recently been introduced.[26] Based on a unique pump–probe beam alignment at obliquely crossed angles, the system provides a low background signal. A key feature of this design is that photothermal noise sources from the optics are eliminated by crossing the pump and probe beams in only the interrogation region.[26] The differential response of the instrument provides a dynamic reserve of 2×10^{-5}, which compares most favorably to the 1×10^{-3}, found in most UV absorbance measurements.

Thermal lens detection has an unusual feature in that the signal is proportional to the power density of the laser beam.[32] This is not true for absorption, fluorescence, or polarimetric measurements. A limitation in thermal lens response occurs with increasing path lengths only several confocal parameters long.[32]

Because indirect thermal lens detection has not yet been evaluated for micro separations, estimating its utility is speculative at present. The limitation could be in focusing the two laser beams to the small flow cell volumes required by these separations.

Recent developments in laser technology may enhance the utility of universal laser-based detection. Semiconductor lasers that are currently being developed may permit frequency-doubling efficiencies of ca. 10% even in the UV.[82] Additionally, semiconductor lasers are inexpensive, they can be tuned over a narrow wavelength range, they generally have lower flicker noise than gas lasers, they can be modulated (pulsed) quite easily, and they are rather easy to package in configurations appropriate for capillary flow cells. These features suggest that the use of semiconductor lasers would have profound implications for almost all of the laser-based techniques discussed, including indirect absorbance, fluorescence, polarimetry, scattering, and indirect detection.

1.4. Summary and Future Developments

We have shown that fundamental measurement science principles, such as the instrument response function, calibration sensitivity, and mass and concentration sensitivity, largely determine the utility of the various laser-based micro separation techniques. For example, the Beer's law path length constraints found in laser-induced fluorescence, direct absorbance measurements, and even laser-based polarimetry may limit the ultimate concentration sensitivity for these techniques. We expect these measurement principles to be an important factor in the future success of these unique detectors as well.

The emergence of new technologies may aid the development of future laser-based detectors. One such example is the microfabrication of integrated optical devices through stereolithography. This new technique permits the fabrication of both optical and electronic components on a single structure, with assembly accomplished through hybrid and surface mount techniques. The recent work with capillaries as waveguides for photon detection of separation processes,[20] as well as improvements in both fiber optic and capillary technology itself, may also improve laser-based detector performance.

The increasing regulatory demands in many disciplines require separation of more and more chemical species with less sample available. Perhaps some evidence of complying with these regulations are the explosive growth in commercially available CE instrumentation over the last 2–3 yr, and the recent emergence of off-the-shelf laser detectors for both HPLC and CE. It is very likely that laser-based micro separation techniques will enjoy further growth in the future, both by continuing as a research tool and by gaining popularity as a routinely used detector.

1.5. Acknowledgments

The authors would like to thank Dr. Donald R. Bobbitt (University of Arkansas) and Dr. Robert A. Weinberger (CE Technologies) for their technical expertise, and The R. W. Johnson Pharmaceutical Institute Department of Scientific Information Resources (Spring House, PA) for their assistance in preparing this manuscript.

1.6. References

1. R. P. W. Scott, in *Liquid Chromatography Detectors*, Elsevier: New York, 8 (1986).

2. D. A. Skoog, *Principles of Instrumental Analysis*, Saunders: Philadelphia, 1 (1985).

3. J. C. Sternberg, in *Adv. in Chromatography*, J. C. Giddings and R. A. Keller, Eds., Dekker: New York, **2**, 251 (1966).

4. D. A. Skoog, *Principles of Instrumental Analysis*, Saunders: Philadelphia, 22–23 (1985).

5. G. L. Long and J. D. Winefordner, *Anal. Chem.* **55** (1983), 712A.

6. E. S. Yeung, in *Microcolumn Separations*, M. Novotny and D. Ishii, Eds., Elsevier: New York, 135–58 (1985).

7. E. S. Yeung, Ed., *Detectors for Liquid Chromatography*, Wiley: New York (1986).

8. D. Ishii and T. Takeuchi, in *Introduction to Microscale High Performance Liquid Chromatography*, D. Ishii, Ed., VCH Publishers: New York, 1–5 (1988).

9. R.P.W. Scott, in *Liquid Chromatography Detectors*, Elsevier: New York, 22–25 (1986).

10. M. Novotny, *Anal. Chem.* **53** (1981) 1294A.

11. V. L. McGuffin and R. N. Zare, in *Chromatography and Separation Chemistry: Advances and Developments,* S. Ahuja, Ed., ACS Symp. Series 297, Amer. Chem. Soc.: Washington, 120–26 (1986).

12. H. Poppe, *Anal. Chim. Acta* **145** (1983) 17–26.

13. E. S. Yeung, *Acc. Chem. Res.* **22** (1989) 125–30.

14. J. C. Giddings, *Unified Separation Science,* Wiley: New York (1991).

15. M. Novotny, *J. Microcol. Sep.* **2**(1) (1990) 7–20.

16. F. J. Yang, Ed., *Microbore Column Chromatography: A Unified Approach to Chromatography,* Dekker: New York (1989).

17. M. J. Sepaniak and E. S. Yeung, *Anal. Chem.* **49** (1977) 1554–57.

18. R. B. Green, *Anal. Chem.* **55** (1983) 20A–32A.

19. C. N. Renn and R. E. Synovec, *Appl. Spectrosc.* **43** (1989) 1393–98.

20. X. Xi and E. S. Yeung, *Anal. Chem.* **62** (1990) 1580–85.

21. Lecture notes from "Practical Capillary Electrophoresis," R. A. Weinberger, CE Technologies, p. X.4 (1991).

22. X. Xi and E. S. Yeung, *Anal. Chem.* **63** (1991) 490–96.

23. J. Pawilszyn, *Anal. Chem.* **60** (1988) 766–73.

24. H. Small and T. E. Miller, *Anal. Chem.* **54** (1982) 462–69.

25. T. Takeuchi and E. S. Yeung, *J. Chromatogr.* **366** (1986) 145–52.

26. S. R. Erskine and D. R. Bobbitt, *Anal. Chem.* **61** (1989) 910–12.

27. H. H. Stuting et al., *LC-GC* **7** (1989) 402–16.

28. Technical note for the ACS ChiraMonitor from Polymer Laboratories Inc., Amherst, MA (1990).

29. Technical note on the IRIS 2000 from Europhor Instruments, Toulouse, France (1991).

30. W. Worthy, *C&EN,* 21–23, Sept. 17, 1990.

31. A. R. Newman, *Anal. Chem.* **63** (1991) 25A–27A.

32. E. S. Yeung, in *Analytical Applications of Lasers,* E. H. Piepmeier, Ed., Wiley: New York, 557–86 (1986).

33. J. N. Harris and N. J. Dovichi, *Anal. Chem.* **52** (1980) 695A–706A.

34. S. Oda and T. Sawanda, *Anal. Chem.* **53** (1981) 471–74.

35. M. D. Morris, in *Detectors for Liquid Chromatography,* E. S. Yeung, Ed., Wiley: New York, 105–47 (1986).

36. E.P.C. Lai et al., *Chromatographia* **15** (1982) 645–49.

37. N. J. Dovichi et al., *Spectrochimica Acta* **43B** (1988) 639–49.

38. M. J. Sepaniak et al., *Anal. Chem.* **56** (1984) 1252–57.

39. K. J. Skogerboe and E. S. Yeung, *Anal. Chem.* **58** (1986) 1014–18.

40. M. Yu and N. J. Dovichi, *Mikrochim. Acta.* **III** (1988) 27–40.

41. V. L. McGuffin, *Proc. Symp. Capillary Chromatogr.,* 6th, P. Heutig: Heidelberg, 800–8 (1985).

42. R. N. Zare, *Science* **226** (1984) 298–303.

43. M. J. Sepaniak and E. S. Yeung, *Anal. Chem.* **49** (1977) 1554–56.

44. E. J. Guthrie et al., *J. Chromatogr. Sci.* **22** (1984) 171–76.

45. M. J. Gordon et al., *Science* **242** (1988) 224–28.

46. H. Drossman et al., *Anal. Chem.* **62** (1990) 900–3.

47. T. J. Edkins and D. C. Shelly, *J. Chromatogr.* **459** (1989) 109–18.

48. V. L. McGuffin and R. N. Zare, *Appl. Spectrosc.* **39** (1985) 847–56 (1985).

49. D. C. Nguyen et al., *J. Opt. Soc. Am. B* **4** (1986) 138–43.

50. T. J. Edkins, Ph.D. Thesis, Stevens Institute of Technology, 114–34 (1990).

51. N. J. Dovichi et al., *Anal. Chem.* **56** (1984) 348–54.

52. S. A. Soper et al., *Anal. Sci.* **5** (1989) 23–29.

53. T. J. Edkins and D. C. Shelly, "Calibration of a Laser-Induced Fluorescence Micro LC Detector," Eastern Analytical Symposium, New York, 1989, paper 137.

54. S. L. Folestad et al., *Anal. Chem.* **54** (1982) 925.

55. M. Albin et al., *Anal. Chem.* **63** (1991) 419.

56. G. G. Lyle and R. E. Lyle, in *Asymmetric Methods of Synthesis,* Academic Press: New York, Vol. 1, 13–27 (1983).

57. S. A. Matlin, in *Chiral Liquid Chromatography,* Chapman and Hall: New York, 16 (1989).

58. D. E. Drayer, *Clin. Pharmacol. Ther.* **40** (1986) 125–33.

59. N. Purdie and K. A. Swallows, *Anal. Chem.* **61** (1989) 78A.

60. E. S. Yeung et al., *Anal. Chem.* **52** (1980) 1399–402.

61. P. D. Rice et al., *Talanta* **36** (1989) 985–88.

62. D. R. Bobbitt and E. S. Yeung, *Anal. Chem.* **56** (1984) 1577–81.

63. R. E. Synovec and E. S. Yeung, *Anal. Chem.* **57** (1985) 2606–10.

64. R. E. Synovec and E. S. Yeung, *J. Chromatogr.* **368** (1986) 85–93.

65. G. D. J. Phillies, *Anal. Chem.* **62** (1990) 1049A–1057A.

66. E. S. Yeung, in *Advances in Chromatography,* J. C. Giddings, Exec. Ed., Dekker: New York, Vol. 23, 1–63 (1984).

67. R. D. Freeman et al., *Appl. Spectrosc.* **42** (1988) 456–60.

68. H. Kolzumi and Y. Suzuki, *J. High Res. Chromatogr. & Chromatogr. Commun.* **10** (1987) 173–76.

69. V. V. Berry, *J. Chromatogr.* **199** (1980) 219–38.

70. V. V. Berry, *J. Chromatogr.* **236** (1982) 279–96.

71. Sj. Van Der Wal and L. R. Snyder, *J. Chromatogr.* **255** (1983) 463–74.

72. H. Binder, *J. Chromatogr.* **189** (1980) 414–20.

73. A. Stolyhwo, H. Colin, and G. Guiochon, *J. Chromatogr.* **265** (1983) 1–18.

74. S. Hoffman et al., *J. High Res. Chromatogr.* **12** (1989) 260–64.

75. W. G. Kuhr and E. S. Yeung, *Anal. Chem.* **60** (1988) 2642–46.

76. E. S. Yeung and W. G. Kuhr, *Anal. Chem.* **63** (1991) 275A–282A.

77. S. Terabe et al., *Anal. Chem.* **56** (1984) 111–13.

78. J. Gorse et al., *J. High Res. Chromatogr.* **11** (1990) 554–59.

79. M. J. Sepaniak et al., *J. High Res. Chromatogr.* **13** (1990) 679–82.

80. D. R. Bobbitt and E. S. Yeung, *Anal. Chem.* **57** (1985) 271–74.

81. Personal communication with D. R. Bobbitt, May 1991.

82. B. Josefsson, *J. Microcol. Sep.* **1** (1989) 116–18.

2

Long-Lived Luminescence Detection in Liquid Chromatography

C. Gooijer, M. Schreurs, and N. H. Velthorst

Free University, Department of General and Analytical Chemistry
de Boelelaan 1083, 1081 HV Amsterdam, the Netherlands

2.1. Introduction

In liquid chromatography (LC), absorption detection in the ultraviolet and visible part of the electromagnetic spectrum is beyond any doubt the most widely applied detection technique. It is almost universal since any compound that has nonzero absorptivity in the spectral range mentioned can be detected. Nevertheless, there are two disadvantages: for trace-level determinations the sensitivity is not always high enough, and, furthermore, in various applications the selectivity of absorption is too low to detect the analytes of interest in the presence of coeluting compounds. This explains why, especially for trace analysis, luminescence and particularly fluorescence spectroscopy have become more and more popular for detection in liquid chromatography.[1]

First of all, luminescence has an inherent high sensitivity. Since it is far more easy to observe a weak light in the dark (as in luminescence) than to detect a little decrease in the intensity of a strong light (as in absorption), it is readily conceived that in fluorescence the attainable detection limits are 2 to 3 decades lower than in absorption detection. In laser-induced fluorescence detection even gain factors of 5 decades have been reported, but it should be realized that such factors are only attainable under exceptional conditions not commonly met in real sample analysis (no chromatographic interferences, analytes, or labeled analytes with excitation maximum at the available laser wavelength).[2,3] Furthermore, fluorescence detection has an inherent selectivity. First, there are two instrumental parameters to be selected, that is, the excitation and the emission wavelength. Second, many compounds do not produce any fluorescence at all. Those compounds, after being excited due to absorption of light, decay to the electronic ground state without any emission of light. As a consequence, in fluorescence detection generally the role of interference is deteriorating much less than in absorption detection. The other consequence, however, is that only a limited number of analytes exhibit native fluorescence so that frequently chemical derivatization has to be involved.[4]

Obviously, fluorescence is the best known and most widely applied representative of luminescence. Nevertheless, it is noted that there are other molecular luminescence methods, such as chemi/bioluminescence, where instead of a lamp a chemical reaction is responsible for the excitation process,[5] and phosphorescence, where the emission of light is based on the transition from the lowest electronic triplet state to the electronic ground state of the molecule.[6] From an analytical-chemistry point of view, the most striking feature of chemi/bioluminescence is that (due to the absence of an excitation source) scattered excitation light does not hinder the detection process. The most important aspect of phosphorescence is its long radiative lifetime: It ranges from 10^{-3} to 10 s, whereas fluorescence lifetimes are in the 10^{-9} to 10^{-7} s region.

Various articles devoted to chemiluminescence detection in liquid chromatography have been reported in the literature.[5,7] They are not discussed here, since the present chapter focuses on long-lived luminescence detection. Within this context it is noted that long-lived luminescence is not limited to molecular phosphorescence of organic compounds. Also some lanthanide ions emit long-lived luminescence in liquid solutions. For practical reasons the lanthanide luminescence will have even more potential in analytical chemistry, because it is observed in the presence of oxygen, whereas phosphorescence of organic compounds in liquid solutions, among other requirements, is only possible if careful solvent deoxygenation is applied, and entrance of oxygen is prevented. Because of its time dependence lanthanide luminescence is extensively utilized for analytical purposes, such as in immunoassays. Especially, europium chelate labels are frequently used, as outlined in a recent review paper written by Diamandis and Christopoulos.[8] However, in this chapter only detection in LC will be considered.

2.2. Room-Temperature Phosphorescence in Liquids

For thoroughly deoxygenated solvents, room-temperature phosphorescence in the liquid state, usually denoted as RTPL, is a rare phenomenon.[6,9] Use has been made of organized media such as micellar solutions to extend the number of compounds that do emit phosphorescence. We will confine ourselves here to normal fluid compositions, frequently used as eluents in liquid chromatography.

At first sight RTPL detection in LC seems to have very little potential, since the number of analytes showing native phosphorescence under such conditions is extremely small. The phosphorescence phenomenon, however, has been successfully utilized in an indirect fashion (i.e., in the sensitized and quenched detection modes). In most applications developed thus far, the phosphorophore is biacetyl (2,3-butanedione) present as a solute in the eluent.

Of course, the standard experimental chromatography setup requires some modifications: The eluent is deoxygenated by purging it with purified nitrogen gas in a specially constructed eluent vessel, and the connections in the liquid flow system are made of stainless steel capillaries, since the commonly used Teflon capillaries are permeable to oxygen.[10,11] Good results have been obtained in a wide range of solvents containing various additional solutes as buffers and ion-pairing reagents.

2.2.1. Sensitized Phosphorescence Detection

In sensitized phosphorescence, generally the analyte absorbs UV radiation but does not emit fluorescence because it decays via the electronic triplet state. In the absence of biacetyl, the subsequent decay step to the ground state takes place nonradiatively, so that phosphorescence is not observable. However, in the presence of a sufficiently high concentration of biacetyl, decay to the ground state may be prevented. Instead, an energy transfer from the excited analyte to biacetyl takes place, and the phosphorescence of biacetyl is monitored. Thus, the crucial point is the efficiency of the energy transfer process. In fact, this efficiency can be quite high, since the lifetime of the triplet state is relatively long.[6,9,10,11]

A simplified energy diagram showing the sensitized phosphorescence pathway is depicted in Figure 2.1. Obviously, the intensity of the sensitized biacetyl phosphorescence signal observed is determined by the following factors:

- The rate of light absorption by the analyte, which is of course proportional to the analyte concentration;
- The efficiency of intersystem crossing to the electronic triplet state, which is an analyte-specific parameter;

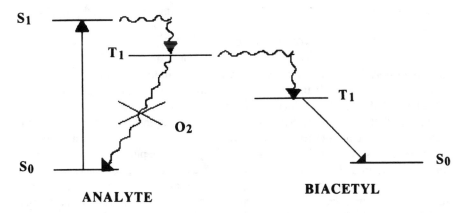

Figure 2.1. Pathway for sensitized biacetyl phosphorescence: The analyte is excited from the singlet electronic ground state S_0, to the excited state S_1; via intersystem crossing it is transferred to the electronic triplet state T_1; then energy transfer to biacetyl takes place, provided that this process is able to compete with the nonradiative decay of the analyte (solvent deoxygenation required!); finally, biacetyl phosphorescence (i.e., the radiative transition from the T_1 to the S_0 state) is observed.

- The efficiency of energy transfer to biacetyl, which is given by the rate of the transfer reaction compared to the rate of the nonradiative decay of the analyte;
- The phosphorescence efficiency of biacetyl under the experimental conditions at hand, which only has a reasonable value for deoxygenated eluents.

The intensivity of the signal is, apart from instrumental factors, given by the product of these four factors and is thus proportional to the analyte concentration, acting as the energy donor.

It will be obvious from Figure 2.1 that only those analytes that have a triplet-state energy higher than that of biacetyl will potentially be detected by sensitized phosphorescence. Furthermore, the analytes need to absorb UV radiation. In other words, UV detection is an alternative, so that it is interesting to compare the chromatograms obtained by both detection techniques.

Finally, it should be realized that in sensitized phosphorescence detection a background phosphorescence signal is observed, since at the excitation wavelength applied, biacetyl itself, present as a solute in the eluent, absorbs a minor amount of radiation and thus provides some direct RTPL. Fortunately the absorptivity of biacetyl is exceptionally low over a large wavelength region so that the background phosphorescence can be kept fairly low. In practice in sensitized phosphorescence the optimum biacetyl concentration allowing an efficient energy transfer on the one hand and a low direct phosphorescence background on the other is on the order of $10^{-4}M$.

Besides, it is noted that in the sensitized phosphorescence detection mode (where the analytes are the energy donors) continuous light sources were normally used; in other words, it is not necessary to utilize time-resolved luminescence detection. In fact, the signal-to-noise ratio in this phosphorescence mode can hardly be improved by invoking time resolution. The signal and the main part of the background have identical time dependencies since they are both coming from biacetyl phosphorescence.

In Figures 2.2 and 2.3 two examples of sensitized phosphorescence detection are shown. The chromatograms in Figure 2.2 are from Arochlor 1221, an industrial mixture of polychlorinated biphenyls. The detection limits are in the low nanogram region. Comparison with the UV detected chromatogram reveals that the ortho-substituted PCBs hardly produce any indirect phosphorescence. Presumably these nonplanar PCBs have too short triplet-state lifetimes to make energy transfer efficient. This feature implies that there is a difference in selectivity between these two detection modes, which is interesting, especially since the planar PCBs are the most toxic.[12] In Figure 2.3, the chromatograms for Halowax 1099, an industrial mixture of polychlorinated naphthalenes, are depicted. Also in this example the difference in selectivity is emphasized. The interesting feature here is that various PCNs (especially the higher chlorinated ones) are not observed in sensitized phosphorescence, because their triplet-state energy is below the triplet-state energy of biacetyl. These particular PCNs can be detected in the quenched phosphorescence mode (see Section 2.2.2).

Finally, it is emphasized that biacetyl itself (and other 1,2-diones) can also be detected via sensitized (biacetyl) phosphorescence; thus, in this case the analyte is

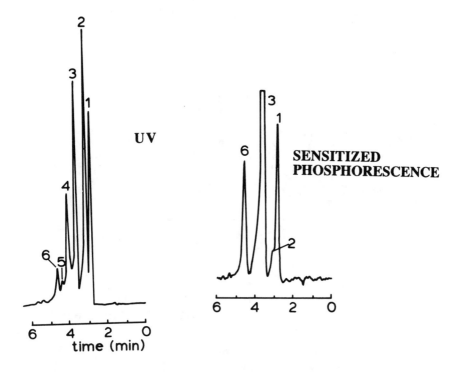

Figure 2.2. Reversed phase chromatograms of an industrial mixture of PCBs obtained by UV absorption detection (244 nm; concentration 50 ppm) and sensitized biacetyl phosphorescence detection (excitation at 265 nm, emission at 520 nm; concentration 10 ppm), showing the selectivity of the latter detection method. Peaks: 1 = biphenyl, 2 = 2-chlorobiphenyl, 3 = 4-chlorobiphenyl, 4 = 2,2′-dichlorobiphenyl, and 6 = 4,4′-dichlorobiphenyl (Ref. 10).

the energy acceptor.[13,14] The low-level determination of biacetyl itself is important in beer, wine, and dairy products. Direct phosphorescence detection of biacetyl is not sensitive enough because of its poor absorptivity. However, indirect excitation can be very efficient via a suitable donor compound so that low concentrations of biacetyl can be detected. A chromatogram is shown in Figure 2.4, applying a substituted naphthalene as the energy donor. Moreover, under these circumstances time-resolved detection is appropriate because there is no phosphorescence background. For this reason in the last example use was made of a commercial luminescence detector equipped with a pulsed source (a pulsed Xe lamp) and a gated photomultiplier. As depicted in Figure 2.5 the pulse width is about 50 μs, and the repetition frequency 50 Hz. For the recording of the chromatogram, we chose a delay time of 100 μs, long enough to avoid the detection of scattered excitation light and possible fluorescence background. Furthermore, the gating time was 900 μs, so that by far the largest part of the biacetyl phosphorescence can be observed. Longer gating times are not appropriate, since after 900 μs hardly any phosphorescence

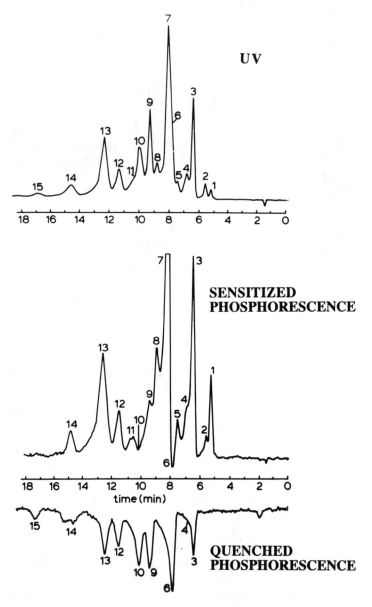

Figure 2.3. Reversed phase chromatograms of an industrial mixture of PCNs (injected amount 1 μg), detected by UV absorption (233 nm), sensitized biacetyl phosphorescence (excitation at 300 nm, emission at 520 nm), and quenched biacetyl phosphorence (excitation at 415 nm, emission at 520 nm). The corresponding peaks are indicated by the numbers 1–15 (Ref. 10).

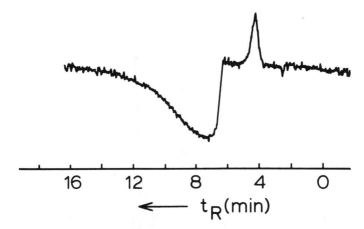

Figure 2.4. Sensitized phosphorescence detection of biacetyl (5 ppb in mobile phase) in the time-resolved mode (see Figure 2.6). Energy donor: 1,5-naphthalenedisulfonic acid disodium salt, (NDSA) $(2 \times 10^{-4}M)$; excitation at 310 nm, emission at 516 nm. The sample was not deoxygenated; the biacetyl peak is clearly separated from the negative oxygen peak eluting from 6.2 to 15 min (Ref. 13).

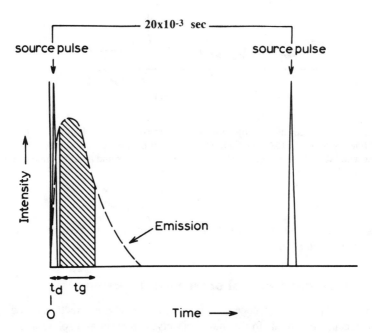

Figure 2.5. Phosphorescence detection in the time-resolved mode. The source pulses are about 50 μs wide; their time difference is 20 msec; t_d is the delay time and t_g the gating time.

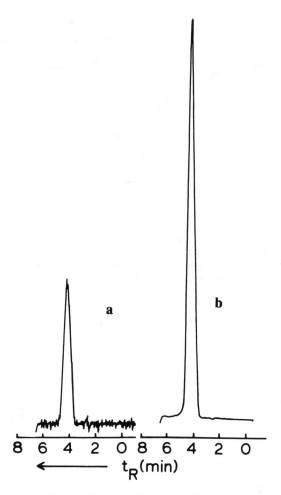

Figure 2.6. The effect of time discrimination in sensitized phosphorescence detection of biacetyl (energy donor NDSA, see Figure 2.4). The chromatogram in (a) was obtained in the "fluorescence" detection mode; in (b) a delay time of 100 μs and a gating time of 900 μsec were applied (Ref. 13).

remains and only noise is recorded. The advantage of time-resolved detection is obvious from Figure 2.6.

2.2.2. Quenched Phosphorescence Detection

In the quenched phosphorescence detection mode, biacetyl is directly excited and its phosphorescence recorded. In this mode biacetyl is present as a solute in the eluent with a concentration of about $10^{-2}M$ (i.e., about a factor 100 higher than in sensitized phosphorescence detection). Hence, in the absence of eluting analytes, a high biacetyl phosphorescence signal is continuously observed. In the presence of certain analytes, however, this strong signal is decreased; that is, negative peaks in the chromatogram are recorded, due to dynamic quenching of phosphorescence.[6,9,10,15]

Dynamic quenching means that the analyte reacts with biacetyl in the excited triplet state, thus prohibiting phosphorescence emission and reducing the net phosphorescence quantum yield. Displacement of the phosphorophore by the analyte plays no role. As a result the lifetime of phosphorescence decreases from τ_0 to τ, and analogously the phosphorescence intensity from I_0 to I, according to the Stern–Volmer equation:

$$\frac{I_0}{I} = 1 + k_q \tau_0 \, [Q].\tag{2.1}$$

In Eq. (2.1), $[Q]$ is the concentration of quenching analyte and k_q the bimolecular quenching rate constant. Of course in fluorescence a similar equation holds. However, in contrast to phosphorescence, in fluorescence dynamic quenching hardly

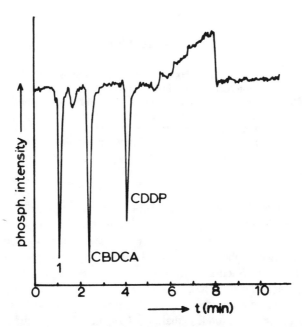

Figure 2.7. Chromatogram of cisplatin, CDDP ($6 \times 10^{-6}M$), and carboplatin, CBDCA ($8 \times 10^{-6}M$), of 0.15 M aqueous NaCl obtained by quenched biacetyl phosphorescence detection. Column: ODS Hypersil, prepared with hexadecyltrimethyl ammonium bromide (HTAB). Mobile phase: water/methanol 99/1 (v/v), pH 5.0 (citrate buffer), $2 \times 10^{-5}M$ HTAB and $1 \times 10^{-2}M$ biacetyl. Excitation at 415 nm, omission at 520 nm. Peak 1: chloride; signal increase from 5.5 to 8.1 min: replacement of HTAB by HTACl (Ref. 16).

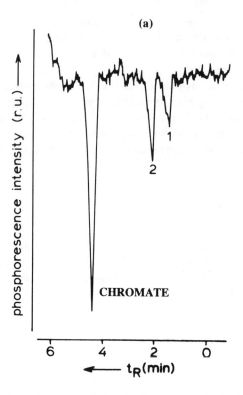

Figure 2.8. (a) Chromatogram of 200 ppb chromate in standard solution obtained by quenched biacetyl phosphorescence detection, in the time-resolved mode (delay time 100 μs; gating time 1.00 ms). Paired-ion chromatography: ODS Spherisorb column (12 cm × 4.6 mm i.d.); mobile phase water/acetonitrile (95/5, v/v), $2 \times 10^{-3}M$ phosphate buffer pH 7.1, $2 \times 10^{-4}M$ tetrabutylammoniumchloride and $5 \times 10^{-3}M$ biacetyl, peaks 1 and 2 quenching impurities (Ref. 17). (b) Chromatograms of surface water, A: blank and B: spiked with $5 \times 10^{-6}M$ chromate; conditions as in (a) (Ref. 17).

plays any role, since the fluorescence lifetime is too low (typically 10–100 ns). In phosphorescence the product $k_q\tau_0$ can be as high as $10^9 \times 10^{-3} = 10^6 M^{-1}$ (an RTPL lifetime of about 1 ms for biacetyl is feasible), which indicates the inherent sensitivity of quenched phosphorescence detection: For $[Q] = 10^{-7}M$ the signal decrease is 10 percent. Such a decrease is easily observed, provided that the noise on the (unquenched) signal is less than 3 percent.

It is emphasized here that the only condition for analytes to be detected in the quenched phosphorescence mode is that their k_q value is high enough. In contrast to sensitized phosphorescence detection, chromophoric properties are not required. Furthermore, from an analytical point of view the mechanism of quenching is not strictly relevant, that is, energy transfer from excited biacetyl to the analyte (the reversed direction compared to the sensitized mode), electron transfer from analyte to excited biacetyl, or reversible hydrogen abstraction.

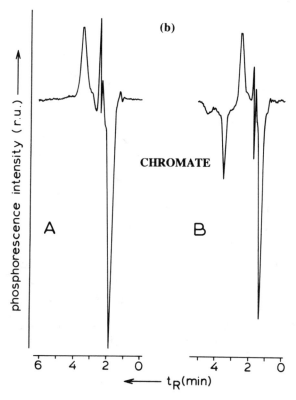

Figure 2.8 (*cont.*)

The implication is that quenched biacetyl phosphorescence detection in LC might be applicable in situations where UV/vis absorption detection fails because of the poor absorptivities of the analytes considered. For this reason it is interesting to explore its applicability in disciplines such as ion chromatography, where in specific cases detection is still a problem. Unfortunately, it is not easy to predict which analytes are efficient quenchers and which are not. Efficient quenching, most probably via electron transfer, has been observed for certain ions occurring in a lower oxidation state, but it is not easy to make generalizations. To quote some examples: nitrite is an efficient quencher whereas nitrate is not; some Pt(II) compounds—for instance, cisplatin and carboplatin—are efficient quenchers, while Pt(IV) complexes are not[16]; however, chromate is an efficient quencher (despite its high oxidation state), while chromium ions are much less efficient.[17] Presumably for chromate there is a low-lying electronic state so that energy-transfer quenching might be operative. Some chromatograms are presented in Figures 2.7 and 2.8.

In the pulsed mode, the Stern–Volmer equation has to be modified (see Figure 2.5). Under normal conditions it is given by[17]

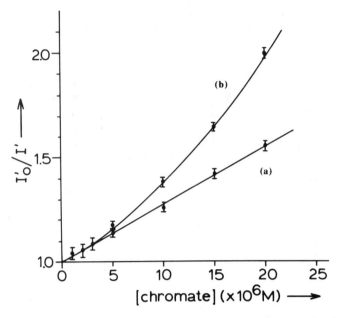

Figure 2.9. I'_0/I' as a function of the chromate concentration calculated from chromatograms (conditions as in Figure 2.8) obtained for delay times of 10 μs (curve a) and 100 μs (curve b) at a gating of 1.00 ms (Ref. 17).

$$\frac{I'_0}{I'} = (1 + k_q\tau_0[Q]) \exp(t_d k_q[Q]). \qquad (2.2)$$

In this modified SV expression the intensity ratio is the product of a linear and an exponential term. The influence of the exponential term decreases when t_d is shortened. This explains why, in the quenched biacetyl phosphorescence detection of chromate (see Figure 2.9), linear calibration curves were obtained in plotting I'_0/I' as a function of chromate concentration if t_d was chosen as short as 10 μs, whereas an obvious deviation of linearity was observed for $t_d = 250$ μs. The general conclusion of this study was that an optimum limit of detection for chromate was achieved (16 ppb or $1.4 \times 10^{-7}M$) for the conditions $t_g = 1.00$ ms and $t_d = 0.100$ ms. For routine analysis requiring a wide linear range, t_d and t_g were set equal to 0.010 and 1.00 ms, respectively. Under the latter experimental settings the LOD was somewhat less favorable, that is, 21 ppb. It is emphasized here that LODs around 20 ppb for chromate are relevant from an analytical point of view. Whereas Cr(III) is an essential trace element to man and plays a role in biochemical processes of aquatic plants and animals, Cr(VI) is an enzymatic poison (which leads to hepatitic and renal damage by exposure) and a suspected carcinogen. Its maximum allowed concentration in drinking water is 50 ppb.

Finally, it is noted that in the literature some experiments have been devoted to immobilized phosphorophores, that is, 1-bromonaphthalene, covalently bonded via an alkyl spacer on glass beads and packed in the luminescence detector cell.[18] Also for this setup, interesting quenched phosphorescence detected chromatograms have been obtained. The effect of scattered excitation light caused by the support could be efficiently suppressed in the time-resolved detection mode by choosing an appropriate delay time. The most obvious advantage of an immobilized phosphorophore compared to the biacetyl system described is that the phosphorophore is not consumed. Unfortunately, also in the immobilized phosphorescence configuration oxygen removal was required to realize phosphorescence detection.

2.3. Lanthanide Luminescence in Liquids

The most obvious disadvantage of the indirect phosphorescence detection methods described in the previous section is the necessity to remove oxygen from the eluent. For this reason it is interesting to explore the detection potential of lanthanide-ion luminescence: Eu(III) and Tb(III) complexes in fluid solutions produce long-lived luminescence, with lifetimes similar to those of RTPL, which is hardly influenced by the presence of dissolved oxygen. Thus, contrary to RTPL, lanthanide luminescence can be utilized in conventional liquid chromatographic instruments without the need to apply eluent deoxygenation and to use stainless steel—instead of Teflon capillaries. It will be shown that, like RTPL, lanthanide luminescence can be applied both in the sensitized and in the quenched detection mode.

In the subsequent sections some spectroscopic features of lanthanide ions are first discussed. Then, the potential of sensitized lanthanide luminescence is outlined. In the presence of oxygen, this detection mode only applies for analytes that form complexes with the lanthanide ion, thus enabling intramolecular energy transfer. Finally, quenched lanthanide luminescence detection in LC is discussed. In addition to analytes causing dynamic quenching of luminescence, in this detection mode analytes that influence the structure of the luminescent lanthanide complex are also monitored.

2.3.1. Spectroscopy of Lanthanides

The absorptivity (ε) of lanthanide ions is very poor.[19] For $EuCl_3$ in water the maximum ε is only 3.06 M^{-1} cm^{-1} (at 394 nm). For $TbCl_3$ in water the situation seems to be somewhat more favorable: The maximum is about 350 M^{-1} cm^{-1}, but, unfortunately, it is achieved at a rather short wavelength, that is, at 225 nm.[20] At this wavelength the spectral output of Xe lamps, commonly used in luminescence detectors, is fairly low. Obviously, especially for Eu(III), the extremely poor absorptivity results in a low emission signal if direct excitation is applied, as in the situation for biacetyl. For this reason generally one has to invoke indirect excitation.

The luminescence lifetime of Eu(III) in water without applying deoxygenation

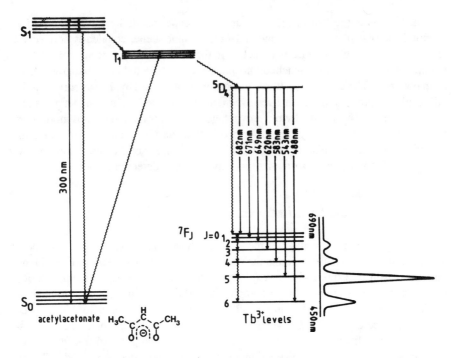

Figure 2.10. Sensitized Tb(III) luminescence induced by excitation of acetylacetonate (compare to Figure 2.1).

was reported to be 106 μs; it is interesting to note that in D_2O it is a factor 20 longer, about 2.27 ms. For Tb(III) the corresponding values are 390 μs in H_2O and about 3.3 ms in D_2O.[20,21] In principle, these values are interesting in the dynamic quenched detection mode, since the sensitivity is directly related to the luminescence lifetime τ_0; see Eq. (2.1). However, because of the poor absorptivities, it is not appropriate to utilize uncomplexed Eu(III) and Tb(III).

For this reason it is useful to consider Eu and Tb complexes, wherein efficient indirect excitation of lanthanides can be realized. Well-known examples are the Tb(III)-acetylacetonate (acac) complex and the Eu(III)-2-thenoyltrifluoroacetate (ttac) complex.[22,23] Presumably these β-diketonates have appropriate triplet-state energy levels to transfer their energy to an "atomic" energy level of the lanthanide ion under study. Besides it is noted that the detailed structure of the β-diketonate is very critical; a particular β-diketonate, able to sensitize Tb(III) luminescence, frequently fails to sensitize Eu(III) luminescence and vice versa. As an example, the indirect excitation in Tb(III) acetylacetonate has been depicted in Figure 2.10. Obviously the emitting level is the $^5D_4$ state; several rather sharp lines are observed in the luminescence spectrum, the most intense being at 545 nm. Luminescence characteristics for Tb-acac and Eu-ttac have been assembled in Table 2.1.

Table 2.1. LUMINESCENCE CHARACTERISTICS
OF THE COMPLEXES[a] Eu-ttac
AND Tb-acac (SEE REF. 23)

	Eu-ttac	Tb-acac
Maximum excitation wavelength (nm)	360	300
Maximum emission wavelength (nm)	614	545
Emitting level of the lanthanide	$^5D_0$	$^5D_4$
Energy of the emitting level (cm^{-1})	17.200	20.500
Luminescence lifetime (ms) in water–ethanol (50/50, v/v)	0.14	0.45

[a] Structures not exactly known.

It is emphasized that the relatively high-resolution character of the Eu(III) and Tb(III) luminescence spectra indicates that the electronic states involved in the transition are more or less shielded from the surroundings; they are hardly influenced if the surroundings are chemically changed. Hence, the spectral line positions are almost independent from complexation; only the relative line intensities are not constant.

Of course, the fact that lanthanide luminescence [in particular for Eu(III)] is observed at such a long wavelength is interesting from an analytical point of view. In principle it is easy to distinguish this luminescence from impurity/background fluorescence, which commonly is observed at much shorter wavelengths.

Finally, it is useful to point out that the lanthanide luminescence quantum yields depend on the chemical surrounding of the lanthanide ion in the complex under consideration. The stronger the complex, the less efficient are the nonradiative deactivation pathways.

2.3.2. The Potential of Lanthanide Luminescence Detection

In view of the discussion of RTPL, it will be obvious from the preceding paragraph that the following points are required in order to estimate the potential of lanthanide luminescence detection in HPLC, both in the sensitized and the quenched mode:

1. Eu(III) and Tb(III) luminescence will be applicable in the sensitized mode, like biacetyl RTPL. Analytes with high triplet yields will be detectable provided their triplet-state energy allows energy transfer to the lanthanide ion under consideration. Direct excitation of the luminophore does not seriously hinder sensitized luminescence detection, since it is very inefficient.

2. In sensitized luminescence the crucial process is the energy transfer to the lanthanide ion, starting from the triplet electronic state of the analyte. Thus the triplet-state lifetime of the analyte should be long enough to make energy transfer at least competitive with nonradiative decay. This explains why for inter-

molecular transfer, occurring if the analyte is not complexed with the lanthanide ion, eluent deoxygenation is required as in sensitized biacetyl RTPL. However, if the analyte forms a complex with the lanthanide, so that energy transfer is an intramolecular process, eluent deoxygenation is not needed.

3. The dynamic quenched detection mode requires complexed Eu(III) or Tb(III) in order to reach sufficiently high luminescence signal levels. Of course, the question arises whether analytes can be found that are able to induce quenching of lanthanide luminescence in such complexes, despite the fact that the ligands protect the lanthanide ion and thus, possibly, prevent the quenching of the luminescent lanthanide. Furthermore, it is evident that not all quenchers known from the quenched biacetyl phosphorescence detection mode will act as quenchers here, the most pronounced example being oxygen itself!

4. It would be interesting to account for the possibility that static quenching is used for analytical purposes. In contrast to dynamic quenching, in static quenching the luminescence quantum yield is not reduced, but the decrease of luminescence is caused by a reduction of the excitation efficiency. It will be observed if the analyte easily displaces a donating ligand of the luminescent lanthanide complex. Analytes that form strong complexes with the lanthanide are expected to be efficient static quenchers.

In the subsequent sections examples of these lanthanide luminescence detection modes will be outlined, although the subject is just starting and far from being fully explored yet. The point to be emphasized now is that the sensitized mode at first sight seems to allow limited application assuming that standard HPLC instrumentation should be applied (without eluent deoxygenation): Only exceptional analytes that form stable complexes and provide appropriate triplet energy levels will be detectable. However, it will be shown that a precolumn derivatization can be applied that greatly widens the detection potential of sensitized lanthanide luminescence. For this reason the subsection on sensitized lanthanide luminescence is divided into two parts: The first is devoted to untreated analytes; the second to derivatized analytes.

2.3.3. Sensitized Lanthanide Luminescence Detection/ Underivatized Analytes

The possibility of sensitized lanthanide luminescence was shown as early as 1965 by Heller and Wasserman.[24] However, it received relatively little attention in the analytical chemistry literature until the beginning of the 1980s. As far as we know, only Wenzel and collaborators have applied sensitized lanthanide luminescence for detection of underivatized analytes in HPLC.

In their pioneering paper on sensitized lanthanide luminescence, Wenzel and co-workers describe some preliminary experiments on mobile phase compatibility using various lanthanide salts in a postcolumn reaction flow.[25] They considered some model analytes, 4,4'-dimethoxybenzophenone (DMB), p-diacetylbenzene, 1-naphthaldehyde, and 1-acetonaphthone. As expected, precautions to avoid quench-

Table 2.2.   RELATIVE PEAK HEIGHTS OF SENSITIZED AND QUENCHED LANTHANIDE LUMINESCENCE[a] FOR DIFFERENT Tb(III) AND Eu(III) SALTS IN PURE METHANOL (SEE REF. 25)

	$TbCl_3$	$Tb(NO_3)_3$	$EuCl_3$	$Eu(NO_3)_3$
4,4'-Dimethoxybenzophenone	100	12	1	0
p-Diacetylbenzene	45	11	5	5
1-Naphthaldehyde	−66	−9	37	9
1-Acetonaphthone	−22	−10	4	3

[a] Indicated by a plus and a minus, respectively.

ing by oxygen were needed for the detection of these compounds, since the energy transfer is intermolecular. Among the four model compounds mentioned, only DMB and p-diacetylbenzene are able to sensitize Tb(III) luminescence. The triplet-state energy of the other two compounds is too low; for these compounds energy transfer in the opposite direction takes place, leading to dynamic quenching of direct Tb(III) luminescence (for comparison, see Section 2.2.2).

DiBella et al. observed that the choice of the lanthanide salt in the postcolumn flow is important: The relative peak height (a positive sign indicates sensitized luminescence, a negative sign quenched luminescence) strongly depends on the anion under consideration (see Table 2.2). Furthermore, an obvious influence of the water content of the mobile phase on the peak intensities was found (see Table 2.3). Fortunately, the deteriorating role of water, being, of course, an essential solvent in reversed-phase liquid chromatography, can be reduced by adding potassium acetate. Nevertheless, we conclude that in general the use of luminescent "free" lanthanide ions in liquid chromatography detection will have only limited potential. To make lanthanide luminescence detection feasible in liquid chromatography, complexed lanthanide ions have to be used.

In a subsequent paper Wenzel and Collette used Eu(III) and Tb(III) for the liquid chromatographic detection of nucleotides and nucleic acids.[26] The lanthanide was applied both in a pre- and postcolumn mode. Obviously, for the base moieties xanthine, guanine, and thiouridine, complexes with Tb(III) are formed, so that intramolecular energy transfer takes place and oxygen plays only a minor role.

Table 2.3.   RELATIVE PEAK HEIGHTS OF SENSITIZED AND QUENCHED Tb(III) LUMINESCENCE FOR $TbCl_3$ IN METHANOL/WATER MIXTURES OF VARYING COMPOSITIONS (SEE REF. 25)

MeOH/water	100:0	95:5	90:10	85:15	80:20
4,4'-Dimethoxybenzophenone	100	15	12	8	7
p-Diacetylbenzene	45	14	15	13	14
1-Naphthaldehyde	−66	−13	−9	−6	−5
1-Acetonaphthone	−22	−2	0	0	0

Other base moieties are not able to produce energy transfer. The applicability of the method was demonstrated for the detection of homologous polynucleotides such as poly-X and poly-G. Furthermore, the possibility of detecting transfer-RNA from *E. coli* was shown using size exclusion chromatography. In the latter example the lanthanide [Tb(III) being more favorable than Eu(III)] was simply added to the mobile phase; the retention for the various tRNA compounds was not altered by the presence of lanthanide in the mobile phase, but a chromatographic separation was not presented by the authors.

Real chromatograms were shown in a subsequent study by Wenzel et al. devoted to the analysis of tetracyclines.[27] In this study sensitized Eu(III) luminescence was used for detection. The applicability of the technique was shown for urine and blood serum samples and for gingival crevice fluid. The tetracyclines used in this investigation are shown in Figure 2.11. Only the postcolumn mode was applied; the flow rates of eluent and postcolumn flow were both 1 mL/min, providing an eventual Eu(III) concentration in the detector cell of $5 \times 10^{-5}M$. Various mobile phase compositions were compared.

It is noted that the authors had to solve a compatibility problem. Under the mobile phase conditions used for the separation of tetracyclines (i.e., a low pH), sensitized Eu(III) luminescence is known to be inefficient. This pH dependence presumably reflects changes of the site of binding; obviously tetracyclines have more possible binding positions for Eu(III), their relative importance being influenced by the pH at hand. The authors solved the compatibility problem by adjusting the pH of the LC effluent by means of a postcolumn flow. Hydroxide and/or oxide formation of Eu(III) (to be expected at pH values above 7) leading to clogging problems was prevented by adding EDTA to the phases containing Eu(III). Lan-

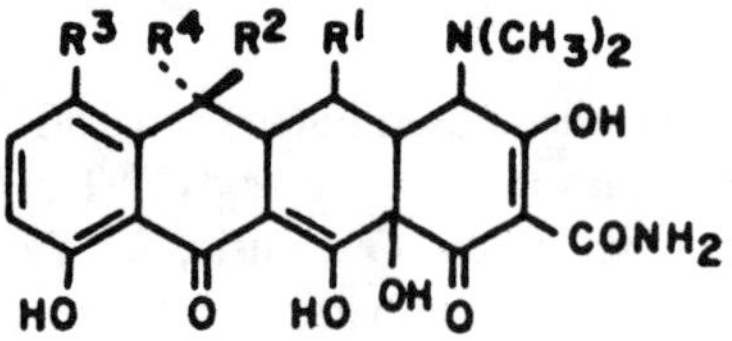

R^1	R^2	R^3	R^4	
H	OH	H	CH$_3$	TETRACYCLINE
H	OH	Cl	CH$_3$	CHLORTETRACYCLINE
OH	H	H	CH$_3$	DOXYCYCLINE
OH	OH	H	CH$_3$	OXYCYCLINE
H	H	H	H	MINOCYCLINE
H	OH	Cl	H	DEMECLOCYCLINE

Figure 2.11.   Tetracyclines investigated in Ref. 27.

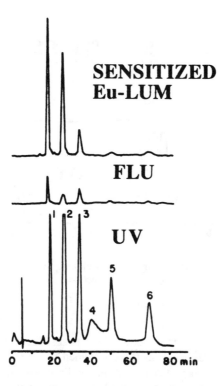

Figure 2.12. Chromatograms of a mixture of tetracyclines $(10^{-5}M)$, oxycycline (1), tetracycline (2), democycline (3), minocycline (4), chlortetracycline (5), and doxycycline (6), obtained by (a) UV absorption at 280 nm; (b) fluorescence detection, excitation at 392 nm, emission at 616 nm; (c) sensitized Eu(III) luminescence, excitation at 392 nm, emission at 616 nm (Ref. 27). For details, see the text.

thanide EDTA complexes are known to be water soluble and stable at pH values as high as 12. The interesting point is that for these complexes additional coordination places for appropriate donors are available; for complexation it is unnecessary to displace the EDTA ligand.

In Figure 2.12 chromatograms for a mixture of tetracyclines at a concentration level of $10^{-5}M$ are presented. The detection modes applied are (a) UV detection at 280 nm; (b) luminescence detection in absence of Eu(III), and (c) luminescence detection in presence of Eu(III). The mobile phase (1 mL/min) was methanol/acetonitrile/0.01M oxalic acid in water pH 2 (1:1.5:5). In Figure 2.12(b) the postcolumn flow was 0.2M ammonium chloride in water (pH 9.0), in Figure 12(c) a solution of Na[Eu(EDTA)].5H$_2$O $(10^{-4}M)$ and ammonium chloride (0.2M) in water (pH 9 using concentrated aqueous ammonia). When the chromatograms in (b) and (c) are compared, it is obvious that for oxycycline and tetracycline [peaks (1) and (2)], sensitized Eu(III) luminescence is much stronger than fluorescence detection. Minocycline (peak 4) is only observable in UV detection and does not produce

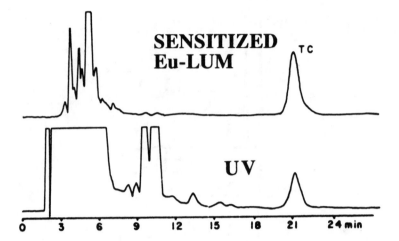

Figure 2.13.   Chromatogram of urine spiked with tetracycline (TC), obtained by UV detection and by sensitized-Eu(III)-luminescence detection; conditions as in Figure 2.12 (see Ref. 27).

any luminescence at all. For the other three peaks direct luminescence detection provides peaks about as weak as those obtained with sensitized Eu(III) luminescence detection. For tetracyclines the detection limit obtained by sensitized Eu(III) luminescence was $5 \times 10^{-7}M$. Figure 2.13 illustrates the selectivity of the sensitized luminescence detection mode by showing chromatograms of urine spiked with tetracycline obtained by using UV detection (at 280 nm) and sensitized Eu(III) luminescence detection. The chromatographic conditions are the same as in Figure 2.12. Obviously, the likelihood that a coeluting peak would have the same requirements for energy transfer to Eu(III) and thus interfere with the detection of tetracyclines is reduced.

2.3.4. Sensitized Lanthanide Luminescence/Derivatized Analytes

The results described in the previous section indicate that sensitized lanthanide luminescence detection has an inherent selectivity.

At first, for nondeoxygenated solvents there will be only few analytes that form stable complexes with Tb(III) or Eu(III) and furthermore have a triplet electronic state enabling energy transfer to the lanthanide ion. Second, owing to the long wavelength of lanthanide luminescence, especially for Eu(III), it can be easily distinguished from impurity and background fluorescence, which is commonly observed at shorter wavelengths. Last, but not least, its time dependence can be involved to separate lanthanide luminescence from background scatter and fluorescence using time-resolved fluorescence.

The most obvious disadvantage of sensitized lanthanide luminescence seems to be its rather limited applicability: Only a few analytes will meet the requirements of producing sensitized luminescence. To extend the applicability range in our group at the Free University in Amsterdam, we have developed a precolumn labeling reaction for thiol-containing analytes.[28–30] Thus far a similar approach has not been published for other functional groups.

It will be shown that development of a derivatization procedure based on eventual monitoring by sensitized lanthanide luminescence is especially appropriate for trace analysis of complex matrices. Compared to existing fluorescence labeling methods, enhancement of selectivity can be obtained. In general, derivatization reactions should meet a number of requirements in order to be applied successfully for detection in liquid chromatography, especially if real samples have to be analyzed.[4] Most important, the reaction should be well defined and not too slow for real samples containing only minor concentrations of analyte in the presence of high amounts of possible interferences. To fulfill this condition, usually the reaction is performed using an excess of labeling reagent. Ideally, the liquid chromatographic performance is sufficient to separate the labeled analytes completely from other eluting peaks. In practice, however, the detection is frequently hindered by interference coming from coeluting sample constituents, excess labeling reagent, and/or reaction side products. For this reason it is appropriate to develop a labeling reaction based on the inherently selective lanthanide luminescence detection technique, despite the fact that in fluorescence detection a wide variety of labeling reactions have already been investigated.

As noted, a derivatization reaction was developed for thiol-containing analytes, denoted as R-SH. The idea is to derivatize the analyte to a product that, in the first place, is able to form a stable complex with Eu(III) or Tb(III) and, in the second place, to sensitize the lanthanide luminescence. Two commercially available derivatization reagents known from fluorescence-labeling reactions have been studied, 4-maleimidylsalicylic acid (denoted 4-MSA) and 4-iodoacetamidesalicylic acid (denoted 4-ISA). These reagents were chosen because the salicylate group has sensitizing properties for Tb(III) luminescence.

It was obvious from batch experiments that 4-MSA is much more favorable than 4-ISA. It can be applied in excess amounts since its unreacted form does not produce any fluorescence nor, in the presence of Tb(III), any sensitized terbium luminescence, either under neutral or extremely alkaline conditions (pH 12). In contrast, unreacted 4-ISA produces both fluorescence and sensitized Tb(III) luminescence. Hence, a surplus of 4-MSA can be utilized in the reaction without any problems, whereas 4-ISA would possibly cause detection interferences.

The reaction between R-SH and 4-MSA, schematically depicted in Scheme 2.1, has been applied in the precolumn mode. For glutathione (GSH) as a model thiol compound, the reaction is completed within 10–15 min in Tris buffer solutions at pH 7.0–7.5, using about fiftyfold excess 4-MSA.

In the first experiments, reversed-phase chromatography was performed with water/acetonitrile (70/30, v/v) at a low pH (i.e., 2.7, HCl added) as the eluent and

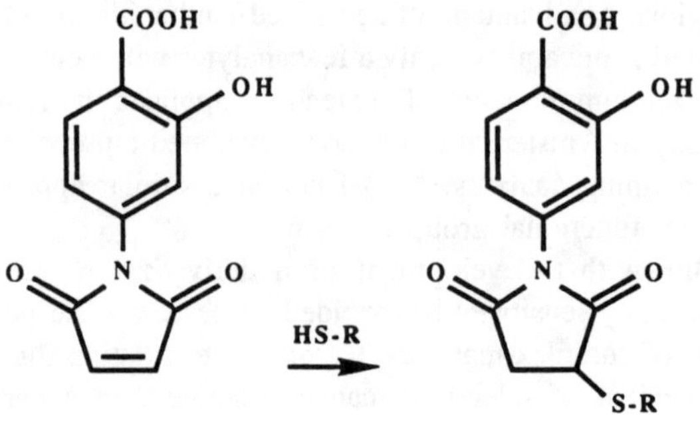

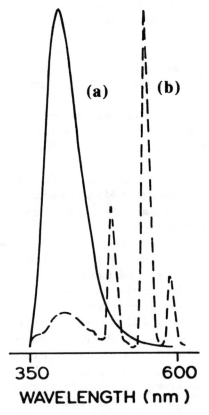

Scheme 2.1. Reaction mechanism between *R*-SH and 4-MSA.

Figure 2.14. (a) Fluorescence spectrum of the reaction product of GSH and 4-MSA; (b) sensitized Tb(III) luminescence spectrum of the same product.

an RP-18 column.[28,29] At this low pH, complexation with Tb(III) is not efficient, so that two separate postcolumn reagent addition lines were needed: one containing $5 \times 10^{-3}M$ Tris buffer (pH 7.5) and a second line containing $2 \times 10^{-3}M$ TbCl$_3$ in water. Buffer and Tb(III) solutions were added separately to prevent hydrolysis of Tb(III) in the postcolumn solutions.

In an alternative approach, use was made of ion pair chromatography by applying water/acetonitrile (75/25 v/v) containing $5 \times 10^{-3}M$ Tris buffer (pH 7.2) and $1 \times 10^{-3}M$ tetrabutylammonium bromide.[30] In this system an RP-18 column was also used. Under these chromatographic circumstances a single postcolumn flow line suffices, such as $1 \times 10^{-3}M$ TbCl$_3$ in water/acetonitrile 50/50 (v/v).

Of course, two detection modes can be utilized to record the chromatograms for the same luminescence detector, such as fluorescence and sensitized Tb(III) luminescence detection (typical emission spectra have been depicted in Figure 2.14). In the case of fluorescence detection no Tb(III) solution is added (so that the eluting analytes are not diluted twice and the terbium ion does not quench the fluorescence), and the excitation and emission wavelengths are set at 302 and 408 nm, respectively. These wavelength settings correspond to the maximum of 4-MSA-thiol fluo-

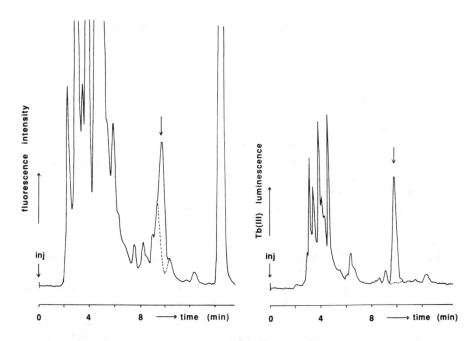

Figure 2.15.   Chromatograms of urine samples (tenfold diluted with $5 \times 10^{-3}M$ Tris buffer pH 7.0 and filtered over a 0.2 μm disposable filter and an octadecyl extraction column) spiked with derivatized $3 \times 10^{-7}M$ N-acetylcysteine (with 4-MSA) detected by fluorescence (left) and by sensitized Tb(III) luminescence (right); the dotted lines represent the unspiked samples.

rescence. To observe sensitized Tb luminescence, the postcolumn flow is needed, and the wavelengths for excitation and emission are 322 and 545 nm, respectively. Apparently, due to the complexation with Tb(III), there is a small shift in the excitation maximum from 302 to 322 nm. In the latter detection mode it is appropriate to utilize time resolution; the delay and gating times were 0.1 and 2.0 ms, respectively.

For standard solutions the difference in detection limits provided by the two detection modes is not extreme. For glutathione (GSH) and N-acetylcysteine (NAC) the Tb luminescence detection mode was found to give a two to three times better signal-to-noise ratio despite the dilution factor caused by the postcolumn flow. The detection limit for GSH in this mode was $2 \times 10^{-8} M$; the calibration curve was linear from 5×10^{-8} to $1 \times 10^{-6} M$.[30] More important is the selectivity of the system. This is illustrated by the chromatograms in Figure 2.15, showing diluted and filtered urine samples spiked with 4-MSA and NAC. As expected, fluorescence detection is not selective enough to observe NAC clearly, since urine contains a lot of fluorescent compounds. The number of compounds that cause efficient sensitizing of Tb(III) luminescence is obviously less. In this detection mode only a little interference is observed in the chromatogram.

2.3.5. Detection Based on Quenched Lanthanide Luminescence

As noted in Section 2.3.1, the free lanthanide ions are not appropriate luminophores to be utilized in dynamically quenched luminescence detection, despite the fact that they have long luminescence lifetimes in liquid solutions. This is due to their poor absorptivities, resulting in a weak and noisy luminescence signal. Evidently a small decrease of such a signal according to the Stern–Volmer equation (see Section 2.2.2) is difficult to detect. In other words, the detection limits achievable are unfavorable. For this reason, lanthanide complexes revealing intense luminescence signals should be involved to realize interesting detectabilities by dynamic quenching lanthanide luminescence detection. Until now, exclusive use has been made of Eu(III) and Tb(III) complexed with ttac and acac, respectively; for convenience they are denoted as Eu-ttac and Tb-acac, since their exact stoichiometry in aqueous solutions is not known.

Eu-ttac and Tb-acac produce intense luminescence signals (showing relatively little noise) because the absorptivities of the ligands guarantee an efficient excitation. Evidently, compounds able to displace ttac and/or acac induce a signal decrease (assuming they are not able to sensitize lanthanide luminescence themselves). Thus, in practice, the negative peaks in the chromatograms are caused either by dynamic quenching or by static quenching of the lanthanide complex under concern. In contrast, for biacetyl phosphorescence only dynamic quenching plays a role (see Section 2.2.2).

The phenomenon of dynamically quenched lanthanide luminescence has already been reported by DiBella et al., who observed negative signals for 1-

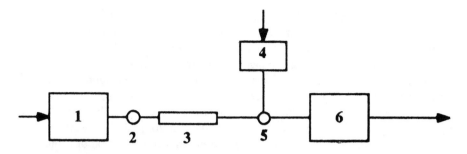

Figure 2.16. Diagram of the LC system used for quenched lanthanide luminescence detection: 1, LC pump, 2, injection valve, 3, analytical column, 4, postcolumn pump, 5, mixing tee, and 6, recorder.

naphthaldehyde and 1-acetonaphthone recording Tb(III) luminescence.[25] The responsible mechanism is energy transfer. In our group we have focused mainly on the detection of inorganic ions. The analysis of chromate in surface and drinking water has been examined,[23] as has been done earlier with quenched biacetyl phosphorescence detection (see Section 2.2.2). A schematic diagram of the LC system is depicted in Figure 2.16. The lanthanide complex was added in a postcolumn flow line. In order to make the setup as simple as possible, we used mixed solutions of ligand and lanthanide salt, for example, $1 \times 10^{-4}M$ $TbCl_3$ and $1 \times 10^{-4}M$ acac in $5mM$ Tris buffer (pH 7.0).

The complex Eu-ttac appeared to be less appropriate than Tb-acac. First of all, in aqueous solutions solubility problems were encountered, and second, its luminescence signal was found to be strongly temperature dependent. The latter point appeared to be a serious problem in our experimental setup. The metal parts are not thermally isolated, and the lamp produces so much heat that a temperature rise of the inlet capillary to the flow cell of more than 10°C in a period of 4 h was observed, associated with a signal decrease of about 50 percent.

In Table 2.4 for a number of inorganic anions the luminescence quenching of Tb-

Table 2.4. INFLUENCE OF INORGANIC ANIONS ON THE LANTHANIDE LUMINESCENCE OF Tb-acac[a] (SEE REF. 23)

NO_3^-, Cl^-, CN^-, SO_3^{2-}, $S_2O_3^{2-}$	No decrease
PO_4^{3-}, CO_3^{2-}, SO_4^{2-}, F^-	Signal decrease probably caused by ligand exchange
NO_2^- (9.9×10^4), CrO_4^{2-} (7.4×10^5), $FeCN_6^{3-}$ (1.3×10^6), $FeCN_6^{4-}$ (9.0×10^5)	Dynamic quenching (quenching constants $k_q\tau_0$ in M^{-1} s)

[a]$1 \times 10^{-4}M$ $TbCl_3$ and $1 \times 10^{-4}M$ acac in 5 mM aqueous Tris buffer (pH 7.0).

acac has been assembled. Three categories can be distinguished: Four anions (among which is chromate) show efficient dynamic quenching; four anions (phosphate being the strongest) induce quenching, presumably via ligand exchange of acac; and the other five do not produce any change of luminescence at all.

Figure 2.17 shows chromatograms of a blank and a spiked sample of surface water, containing $1 \times 10^{-6}M$ chromate. It should be noted that positive signals are observed since the reversed luminescence intensity is recorded. The two peaks in the chromatogram arise from hydrogencarbonate (4 min), present in water at high concentrations, and sulfate (about 9 min), present in a concentration of about $4 \times$

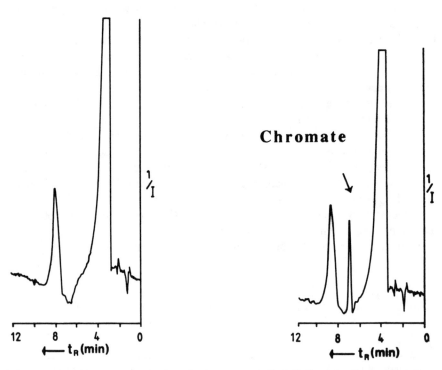

Figure 2.17. Chromatograms of surface water, unspiked (left) and spiked with $1.0 \times 10^{-6}M$ chromate (right) detected by quenched Tb(III) luminescence (time-resolved mode). Column: RP18 (15 cm $\times$ 3.1 mm); eluent: water/acetonitrile (90/10, v/v) containing $5 \times 10^{-4}M$ TBABr and $5 \times 10^{-3}M$ Tris buffer (pH 7.0); postcolumn solution: eluent containing additionally $1 \times 10^{-4}M$ Tb (III) and $1 \times 10^{-4}M$ acac (see Ref. 23).

$10^{-4}M$; both are probably able to form complexes with Tb(III). Nevertheless, chromate can be detected very well, the detection limit being $1 \times 10^{-7}M$ (13 ppb).

2.4. Concluding Remarks

The results described in this chapter clearly illustrate the potential of long-lived luminescence detection in liquid chromatography. It is not only the possibility of time discrimination to improve the signal-to-noise ratio that makes long-lived luminescence worthwhile for detection purposes. Maybe even more important is its selectivity, inherent to the fact that long-lived luminescence in fluid solutions is a rare phenomenon.

Considering the results reported for room-temperature phosphorescence in liquids, the selectivity of sensitized (and quenched) biacetyl phosphorescence has been clearly illustrated for mixtures of polychlorinated biphenyls. The nonplanar compounds hardly produce any signal, whereas the toxic planar PCBs are clearly seen. For mixtures of polychlorinated naphthalenes, both sensitized and quenched phosphorescence detection can be applied, providing complementary information. Of course, the detection of α-diketones (like biacetyl) in beer samples is an interesting but rather exceptional application of sensitized RTPL, since it is based on the phosphoric properties of the analytes under consideration. Quenched biacetyl phosphorescence detection seems to be especially interesting for analytes with poor chromophoric properties, as is frequently encountered in ion chromatography.

The potential of lanthanide luminescence has not been fully explored yet. Nevertheless, the results available are very interesting. Especially the exceptional selectivity of sensitized lanthanide luminescence has to be emphasized: only analytes that form stable complexes with Eu(III) or Tb(III) and furthermore have an appropriate T_1-state energy are observed. This unique selectivity opens the way to make use of derivatization procedures directed at the eventual observation of lanthanide luminescence, which is of particular importance if fluorescence detection is hindered by interferences. The results presented for a number of thiols are quite promising. Quenched lanthanide luminescence suffers from the drawback that not exclusively dynamic quenching processes play a role, as in quenched RTPL. A decrease of luminescence intensity is also observed for eluting compounds that affect the structure of the luminescent lanthanide complex. More experiments have to be done to show the analytical potential of this detection mode.

Finally, it should be realized that the application of phosphorescence detection is strongly hindered by the fact that it is not performed with standard chromatographic instrumentation. It requires careful eluent deoxygenation and stainless-steel interconnections instead of Teflon capillaries. This drawback does not apply for lanthanide luminescence detection. Standard chromatographic instrumentation can be applied, and detectors with a pulsed Xe lamp and a gated photomultiplier are commercially available. Thus, lanthanide luminescence has a real, practical potential.

2.5. References

1. U. A. Th. Brinkman, G. J. de Jong, and C. Gooijer, *Pure & Appl. Chem.* **59** (1987) 625.

2. M. C. Roach and M. D. Harmony, *Anal. Chem.* **59** (1987) 411.

3. C. M. B. van den Beld, H. Lingeman, G. J. van Ringen, U. R. Tjaden, and J. van der Greef, *Anal. Chim. Acta* **205** (1988) 15.

4. U. A. Th. Brinkman, *Chromatographia* **24** (1987) 190.

5. J. W. Birks, Ed., *Chemiluminescence and photochemical reaction detection in chromatography,* VCH: Weinheim, New York (1989).

6. C. Gooijer, R. A. Baumann, and N. H. Velthorst, *Prog. Anal. Spectrosc.* **10** (1987) 573.

7. M. L. Grayeski, *Anal. Chem.* **59** (1987) 1243.

8. E. P. Diamandis and T. K. Christopoulos, *Anal. Chem.* **62** (1990) 1149A.

9. R. J. Hurtubise, *Phosphorimetry* VCH: New York, Chapter 9 (1990).

10. J. J. Donkerbroek, N.J.R. van Eikema Hommes, C. Gooijer, N. H. Velthorst, and R. W. Frei, *J. Chromatogr.* **255** (1983) 581.

11. C. Gooijer, N. H. Velthorst, and R. W. Frei, *Trends in Anal. Chem.* **3** (1984) 259.

12. V. A. McFarland and J. U. Clarke, Environmental Health Perspectives **81** (1989) 225–239.

13. R. A. Baumann, C. Gooijer, N. H. Velthorst, and R. W. Frei, *Anal. Chem.* **57** (1985) 1815.

14. R. A. Baumann, C. Gooijer, N. H. Velthorst, R. W. Frei, J. Strating, L. C. Verhagen, and R. C. Veldhuyzen-Doorduin, *Intern. J. Environm. Anal. Chem.* **25** (1986) 195.

15. J. J. Donkerbroek, A. C. Veltkamp, C. Gooijer, N. H. Velthorst, and R. W. Frei, *Anal. Chem.* **55** (1983) 1886.

16. C. Gooijer, A. C. Veltkamp, R. A. Baumann, N. H. Velthorst, R. W. Frei, and W.J.H. van der Vijgh, *J. Chromatogr.* **312** (1984) 337.

17. R. A. Baumann, M. Schreurs, C. Gooijer, N. H. Velthorst, and R. W. Frei, *Can. J. Chem.* **65** (1987) 965.

18. R. A. Baumann, C. Gooijer, N. H. Velthorst, R. W. Frei, I. Aichinger, and G. Gübitz, *Anal. Chem.* **60** (1988) 1237.

19. W. T. Carnall, "The Absorption and Fluorescence Spectra of Rare Earth Ions in Solution," in *Handbook on the Physics and Chemistry of Rare Earths,* Vol. 3, K. A. Gschneider and L. Eyring, Eds., North-Holland: Amsterdam (1979).

20. R. A. Baumann, D. A. Kamminga, H. Derlagen, C. Gooijer, N. H. Velthorst, and R. W. Frei, *J. Chromatogr.* **439** (1988) 165.

21. J. L. Kropp and M. W. Windsor, *J. Chem. Phys.* **42** (1965) 1599.

22. W. R. Dawson, J. L. Kropp, and M. Windsor, *J. Chem. Phys.* **45** (1966) 2410.

23. M. Schreurs, G. W. Somsen, C. Gooijer, N. H. Velthorst, and R. W. Frei, *J. Chromatogr.* **482** (1989) 351.

24. A. Heller and E. Wasserman, *J. Chem. Phys.* **42** (1965) 949.

25. E. E. DiBella, J. B. Weissman, M. J. Joseph, J. R. Schultz, and T. J. Wenzel, *J. Chromatogr.* **328** (1985) 101.

26. T. J. Wenzel and L. M. Collette, *J. Chromatogr.* **436** (1988) 299.

27. T. J. Wenzel, L. M. Collette, D. T. Dahlen, S. M. Hendrickson, and L. W. Yarmeloff, *J. Chromatogr.* **433** (1988) 149.

28. M. Schreurs, C. Gooijer, and N. H. Velthorst, *Anal. Chem.* **62** (1990) 2053.

29. M. Schreurs, C. Gooijer, and N. H. Velthorst, *Fresenius J. Anal. Chem.* **339** (1991) 499.

30. M. Schreurs, L. Hellendoorn, C. Gooijer, and N. H. Velthorst, *J. Chromatogr.* **552** (1991) 625.

3

Chemiluminescent Detection in High-Performance Liquid Chromatography

David S. Hage

Department of Chemistry
University of Nebraska
Lincoln, Nebraska 68588-0304

3.1. Introduction

Chemiluminescence may be defined as the emission of light as the result of a chemical reaction. In recent years this phenomenon has been of increasing importance as a tool for detection in high-performance liquid chromatography (HPLC). Advantages of chemiluminescent detection are its ability to monitor low concentrations, its selectivity, and its ease of use. This chapter will discuss the basic principles behind chemiluminescence and how it is used for detection in HPLC. Applications of this method will be examined, as well as the advantages of chemiluminescent detection versus other techniques.

3.2. Principles of Chemiluminescence

The basic processes involved in chemiluminescence are shown in Figure 3.1. This involves an exothermic reaction, in which reactants combine to form an excited-state product. This product later relaxes to its ground state by releasing its excess energy. Normally such relaxation occurs through the emission of heat or by collisions with other molecules. However, in some chemical systems a significant fraction of the excited state product also relaxes through the emission of light. It is this light that is measured in chemiluminescent detection.[1]

Chemiluminescence occurs in a number of biological and synthetic chemical systems. Nearly all of these reactions involve an oxidative process based on oxygen, hydrogen peroxide, or oxygen-related metabolites.[1] They also tend to involve reac-

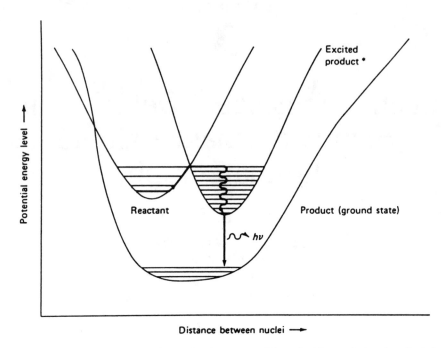

Figure 3.1.   Basic principles of chemiluminescence. Adapted with permission from Ref. 1.

tants or intermediates with high internal energies, such as those with strained ring structures or peroxide groups.[2]

Note that the processes shown in Figure 3.1 are very similar to those involved in fluorescence. The key difference between chemiluminescence and fluorescence is the way in which the excited-state intermediate is produced. In chemiluminescence this is produced by a chemical reaction, but in fluorescence it is produced by the absorption of light. One advantage of this is that chemiluminescence detectors do not need an excitation light source in order for a signal to be produced. This is useful since background due to the excitation light is one of the major obstacles in obtaining low limits of detection with fluorescence monitors. This makes chemiluminescent detection inherently more sensitive than fluorescence. For example, peroxyoxalate-based chemiluminescent detection can be used to monitor some fluorescent compounds at levels 2–3 orders of magnitude lower than can be determined using standard fluorescence measurements.[3] Because of its low background, limits of detection obtained with chemiluminescence are frequently in the femtomole to attomole range (10^{-15} to 10^{-18} moles).[2,3]

3.3. Systems for Chemiluminescent Detection in HPLC

A typical system used for chemiluminescent detection in HPLC is shown in Figure 3.2. This system usually involves combining column eluent with some postcolumn

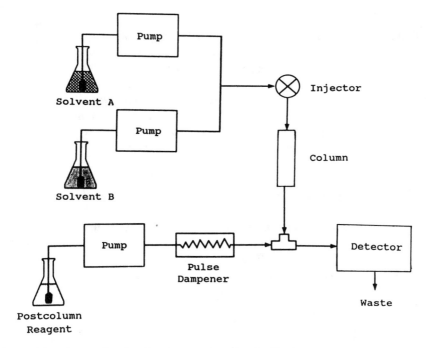

Figure 3.2. System for chemiluminescent detection in HPLC.

reagent that initiates the desired chemiluminescent reaction. The light produced by this reaction soon reaches a maximum and then begins to decline in intensity as reagent or analyte is consumed. To detect light emitted by this reaction, the analyte and reagent mixture is passed through a flow-through detector containing a sensitive photomultiplier tube. Since the signal produced by the chemiluminescent reaction is time dependent, care must be taken to design this system so that as much signal is produced in the detector as possible. Failure to do this can result in a significant fraction of the signal being produced before or after the detector, resulting in a loss of sensitivity.[2]

The amount of light produced in the detector can be adjusted by varying the rate of the chemiluminescent reaction, the delay between the mixer and the detector, and the time that analyte spends in the detector flow cell. For many chemiluminescent systems, the rate of reaction can be controlled by changing the pH or reagent concentrations. The delay can be controlled by adjusting the eluent and reagent flow rates or by varying the volume of tubing between the mixer and the detector. The time the analyte spends in the detector can similarly be controlled by changing the flow rates of the eluent and reagents or by varying the volume of the detector flow cell. Other items important in determining signal intensity include the efficiency of light production, the efficiency of light collection, and the variation in background signal. The efficiency of light production depends on the quantum yield of the particular chemiluminescent reaction being used. This can sometimes be improved by using a more efficient chemiluminescent system or by placing additives in the postcolumn reagent (e.g., surfactants or organic solvents) that improve the quantum light yield of the reaction already in use.[1] The extent of light collection can be improved by placing the detector flow cell close to the photomultiplier tube or by using special reflection systems, such as an ellipsoidal mirror, for better light collection.[2] Fluctuations in the background signal due to pump pulsations can be minimized by using pulse dampeners. Some chemiluminescent reactions also require the use of high-purity solvents and reagents in order to minimize the background signal.[2] By properly controlling these various parameters, limits of detection in the subattomole range are theoretically obtainable.[4]

3.4. Chemiluminescent Reactions and Applications

A number of biological and synthetic chemical systems have been used for detection in HPLC. The following section will discuss the chemical reactions involved in these systems and will examine how these reactions are used for HPLC detection. A review of the applications reported for each of these systems will also be presented.

3.4.1. Bioluminescence

The production of light by a biological system is referred to as *bioluminescence*. Bioluminescent reactions involve the enzyme-catalyzed oxidation of a substrate to produce an excited-state intermediate. The enzyme responsible for this process is known as a *luciferase*, and its substrate is known as a *luciferin*. Many types of luciferases and luciferins exist in nature, occurring in a variety of different organisms.[1] These systems are of interest in analytical applications because of their high light yields, which allow low limits of detection to be obtained with their use.

One of the most familiar examples of bioluminescence is the production of light by the firefly. The reaction involved in this process is shown in Figure 3.3. In this system, the enzyme firefly luciferase catalyzes the oxidation of firefly luciferin to form oxyluciferin and light. Oxygen, adenosine triphosphate (ATP), and magnesium ions are also required for this reaction. This process is one of the most efficient chemiluminescent reactions known. In some cases quantum yields of up to 90% have been reported.[3] The light emitted by this reaction has its greatest intensity in the visible range, with a wavelength maximum at 562 nm. Besides oxyluciferin and light, other products of the reaction are adenosine monophosphate (AMP), magnesium pyrophosphate, carbon dioxide and water.[1,3]

One application of the firefly reaction has been its use for the detection of creatinine kinase (CK) isoenzymes. The determination of these isoenzymes is important in clinical chemistry for the diagnosis of heart attacks and various muscle disorders. Bostick and co-workers detected these isoenzymes as they eluted from an ion-exchange column by using a detection scheme that coupled the enzymatic reaction of creatine kinase with the firefly luciferase system. In the first step of this process, column eluent was combined with a postcolumn reagent containing

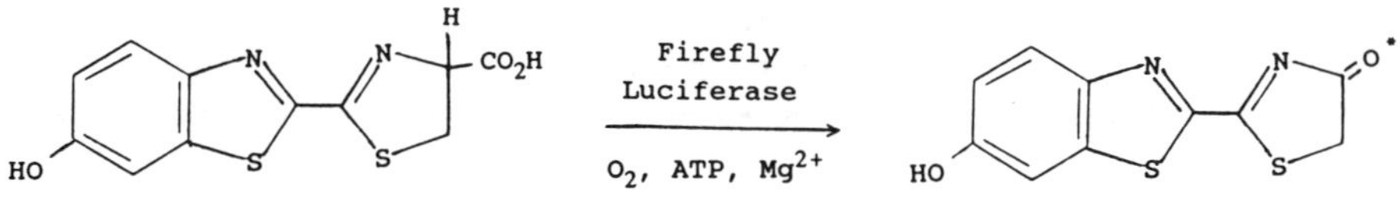

Firefly Luciferin   Oxyluciferin•

Oxyluciferin• ⟶ Oxyluciferin + Light

Figure 3.3.   Firefly bioluminescence.

creatine phosphate and adenosine diphosphate (ADP). As CK isoenzymes eluted from the column, they catalyzed the following reaction:

$$\text{Creatine phosphate} + \text{ADP} \xrightarrow{\text{CK}} \text{Creatine} + \text{ATP}. \qquad (3.1)$$

In the second step of the detection scheme, ATP generated by the CK reaction was combined with firefly luciferin, luciferase, magnesium ions, and oxygen. The latter reactants were added as part of the postcolumn reagent. This resulted in the production of light whenever ATP was generated by CK, providing a signal proportional to the activity of isoenzymes eluting from the column.[5]

Another common type of bioluminescence is that found in numerous deep-sea fish. Light is produced in this case by bacteria that live symbiotically with these organisms. The basic process believed to occur in this type of bioluminescence is shown in Figure 3.4. In this reaction, reduced flavin mononucleotide (FMNH$_2$) is enzymatically oxidized to form an excited molecule of flavin mononucleotide (FMN).[3] Oxygen and a long-chain aliphatic aldehyde (C$_8$–C$_{14}$) are required for this process. The excited-state FMN produced goes to its ground state by emitting light with a quantum yield of about 10% and a maximum emission wavelength between 470 and 505 nm.[1]

Bacterial bioluminescence has been used for the postcolumn detection of bile acids in HPLC. This was performed by Arisue et al. using a scheme that coupled the enzymatic reaction of bacterial luciferase with that of 3-α-hydroxysteroid-NAD oxidoreductase. Oxidoreductase was used to catalyze the reaction of bile acids with nicotinamide-adenine dinucleotide (NAD) to form the reduced form of NAD, or NADH. The NADH formed was then further reacted with FMN in the postcolumn reagent to reform NAD and FMNH$_2$:

$$\text{FMNH}_2 \; + \; \text{RCHO} \; + \; \text{O}_2 \xrightarrow{\quad \substack{\text{Bacterial} \\ \text{Luciferase}} \quad} \text{FMN}^* \; + \; \text{RCOOH} \; + \; \text{H}_2\text{O}$$

$$\text{FMN}^* \xrightarrow{\qquad\qquad} \text{FMN} \; + \; \text{Light}$$

Figure 3.4.   Bacterial bioluminescence.

$$2\text{NADH} + \text{FMN} \xrightarrow{\qquad\qquad} 2\text{NAD} + \text{FMNH}_2. \tag{3.2}$$

The FMNH_2 generated was next used by bacterial luciferase to produce light, according to the reactions given in Figure 3.4. This resulted in a signal proportional to the amount of bile acid eluting from the column.[6]

3.4.2. Peroxyoxalate System

The peroxyoxalate system is one of the most common chemiluminescent reactions used for detection in HPLC.[2,7] The reactions involved in this system are shown in Figure 3.5. First, an aryl oxalate ester and hydrogen peroxide react to form a high-energy intermediate, 1,2-dioxetanedione. In the presence of certain fluorescent

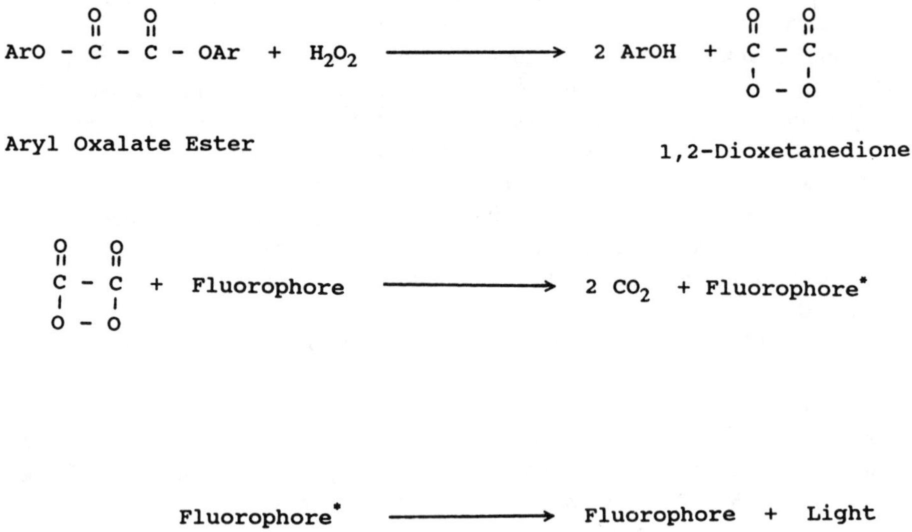

Aryl Oxalate Ester

1,2-Dioxetanedione

$$\text{Fluorophore}^* \xrightarrow{\qquad\qquad} \text{Fluorophore} \; + \; \text{Light}$$

Figure 3.5.   Peroxyoxalate chemiluminescence.

compounds, this intermediate will transfer some of its energy to a neighboring fluorophore. The excited-state fluorophore than releases light as it goes back to its ground state. The wavelengths of light emitted are generally characteristic of the first excited singlet state of the fluorophore.[3] Quantum yields for this reaction range from 1 to 23%.[8]

One advantage of the peroxyoxalate reaction in HPLC detection is that it has one of the highest light yields reported in a synthetic chemiluminescent system. This allows it to be used to determine small quantities of analyte. A second advantage of this reaction is that it can be used to monitor a wide variety of fluorescent compounds without requiring the use of an excitation light source. As already discussed, this results in a response with little or no background, allowing limits of detection to be obtained that are several orders of magnitude lower than those measured using standard fluorescence monitors.[3]

The two most common aryl oxalates used in this reaction are bis(2,4,6-trichlorophenyl)oxalate, or TCPO, and bis(2-nitrophenyl)oxalate, or 2-NPO. For optimum sensitivity, these are mixed with hydrogen peroxide and fluorophore immediately before the flow cell of the detector. Typical concentrations used in the detection of fluorophores are 1 to 10 mM oxalate ester and 10 to 500 mM hydrogen peroxide.[2]

The peroxyoxalate system may be used in HPLC to detect (1) compounds that are naturally fluorescent, (2) compounds labeled with fluorescent tags, (3) compounds that can be coupled with enzymatic reactions that produce hydrogen peroxide, or (4) nonfluorescent compounds that affect the efficiency of light production by the peroxyoxalate system. Examples of naturally fluorescent compounds that have been determined in HPLC using peroxyoxalate chemiluminescence include polycyclic aromatic hydrocarbons,[9,10] polycyclic aromatic amines,[11] and nitropolycyclic aromatic hydrocarbons.[12] Each of these examples used a reversed-phase column and a postcolumn reagent containing TCPO and hydrogen peroxide.[9-12] The determination of nitropolycyclic aromatic hydrocarbons also required prior treatment of the analytes using a zinc reduction column.[12] The limits of detection reported for these systems were typically in the picomole or picogram range.[9-12]

The most popular use of peroxyoxalate chemiluminescence in HPLC has been in the detection of compounds derivatized with fluorescent labels. Labels that may be detected by the peroxyoxalate system have included dansyl chloride (DNP), o-phthaldehyde (OPA), fluorescamine, 3-aminoperylene, and various coumarins.[2,7,13] One application of fluorescent labels and peroxyoxalate chemiluminescence has been in the detection of derivatized amino acids separated by HPLC. Dansyl chloride is generally used as the label in these studies, but work with o-phthaldehyde and other labels has also been reported.[7] An example of a chromatogram obtained with this approach is shown in Figure 3.6. Such separations have been reported for both isocratic and gradient systems, with limits of detection ranging from 0.3 to 10 femtomoles per amino acid. A detailed summary of the chromatographic conditions used in these and related separations has been presented by Imai.[7]

Other compounds determined using fluorescent labels and peroxyoxalate chem-

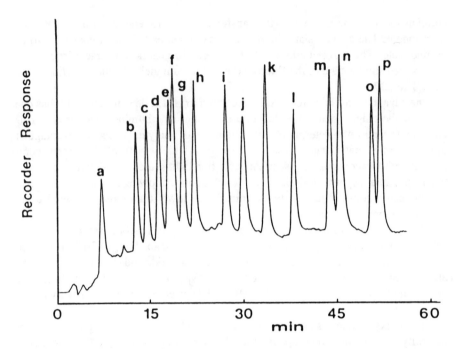

Figure 3.6. Determination of dansylated amino acids using HPLC and peroxyoxalate chemiluminescence. Peak identification: a, Asp; b, Asn; c, Gln; d, Ser; e, Arg; f, Thr; g, Gly; h, Ala; i, Pro; j, Lys; k, Val; l, Met; m, Ile; n, Leu; o, Trp; p, Phe. Amount of each amino acid injected, 50 femtomoles. (Reproduced with permission from Ref. 7.)

iluminescence have included catecholamines,[14,15] estradiol,[16] 3-α-ketocortico-steroids,[17] secoverine,[18] secondary amine-containing drugs,[19] bradykinin,[20] and carboxylic acids.[13,21] In each of these cases, limits of detection in the femtomole or picogram range have been obtained.[2]

Direct detection of a nonfluorescent compound may be performed by using the analyte as a substrate for an enzymatic reaction that generates hydrogen peroxide. The hydrogen peroxide produced is then reacted with an oxalate ester and excess fluorophore to produce light. This approach has been used by Honda et al. in the detection of acetylcholine and choline separated by a reversed-phase column.[22] In this work, an immobilized enzyme reactor containing choline oxidase and acetyl-choline esterase was used to generate hydrogen peroxide as analytes eluted from the column. Eluent leaving the enzyme reactor was then combined with a postcolumn reagent containing TCPO and perylene. The limit of detection for both cholines was 1 picomole per injection.[22]

Van Zoonen and co-workers detected acetylcholine and choline using this same enzymatic system along with a special flow cell containing an immobilized fluo-rophore (3-aminofluoranthene) and solid TCPO.[23] The design of this system is illustrated in Figure 3.7. By using different immobilized enzyme reactors, the same

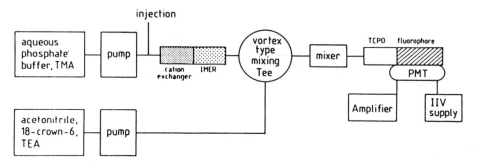

Figure 3.7. Immobilized enzyme reactor for the determination of acetylcholine and choline using HPLC and peroxyoxalate chemiluminescence. (Reproduced with permission from Ref. 23.)

approach has been used in the detection of glucose[24] and L-amino acids.[25] For the determination of glucose, the immobilized enzyme was glucose oxidase, and the oxalate ester was solid TCPO.[24] For the determination of L-amino acids, the immobilized enzyme was L-amino acid oxidase, and the oxalate ester was 2-NPO added as part of a postcolumn reagent.[25] Limits of detection in each case were in the picomole to subpicomole range.[23-25]

Detection of some nonfluorescent compounds with no derivatization can also be achieved using the peroxyoxalate system. This was demonstrated in work by Capomacchia et al.[26] In their studies with bis(2,4-dinitrophenyl)oxalate, they found that the nonfluorescent compounds ouabain and urea produced an enhancement of peroxyoxalate chemiluminescence. The degree of enhancement was approximately 1000-fold less than that obtained with fluorophores; however, the effect was strong enough to allow both compounds to be detected at levels of only 2 to 20 picomoles per injection. Although the exact mechanism behind this enhancement is not currently known, it does represent an interesting alternative for detection using the peroxyoxalate system.[2]

3.4.3. Luminol

The chemiluminescent reaction of luminol (5-amino-2,3-dihydrophthalazine-1,4-dione) is another nonenzymatic system commonly used for detection in HPLC. The steps involved in luminol chemiluminescence are shown in Figure 3.8. The basis of this reaction is the oxidation of luminol to form an excited molecule of 3-aminophthalate, along with nitrogen and water. The 3-aminophthalate then relaxes to its ground state by emission of light with a wavelength maximum at 425 nm.[1,3] The efficiency of this reaction is approximately 1%.[3]

The oxidant for the luminol reaction is usually hydrogen peroxide in the presence of basic solution (pH 10–13) and a catalyst. A variety of catalysts may be used to promote luminol chemiluminescence. Examples include a number of transition metal ions (e.g., Fe^{2+}, Fe^{3+}, Co^{2+}, Ni^{2+}, and Cu^{2+}), as well as transition metal

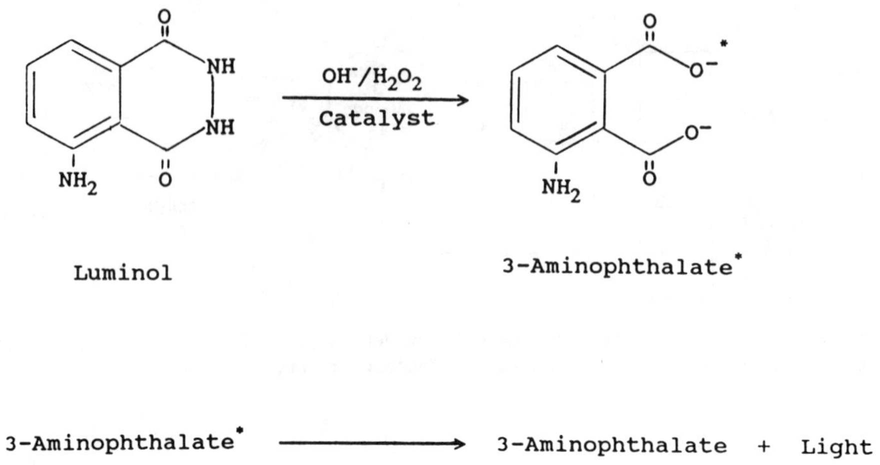

Figure 3.8. Luminol chemiluminescence.

complexes [e.g., $Fe(CN)_6^{3-}$ and $Cu(NH_3)_6^{2+}$]. The reaction is also catalyzed by porphyrins and heme-containing proteins. Heme-containing proteins that catalyze luminol chemiluminescence include hemoglobin, myoglobin, catalase, cytochrome c, and numerous peroxidases.[1]

Luminol chemiluminescence can be used for detection in HPLC in a number of different ways. Applications include the detection of (1) compounds labeled with luminol derivatives, (2) compounds that catalyze luminol chemiluminescence, (3) compounds that inhibit metal-ion-catalyzed luminol chemiluminescence, (4) compounds consumed by enzymatic or nonenzymatic reactions that produce hydrogen peroxide, and (5) hydrogen peroxide–related agents.

The luminol derivative typically used in labeling compounds for HPLC detection is N-(4-aminobutyl)-N-ethylisoluminol, or ABEI. This was used by Kawasaki and co-workers in the detection of bile acids, free fatty acids, primary amines and secondary amines. This work used a reversed-phase column and a postcolumn reagent containing hydrogen peroxide and a potassium hexacyanoferrate catalyst. Using this system in the separation of bile acids, a limit of detection of 20 femtomoles was obtained for cholic acid.[27] In similar work, Yuki et al. used ABEI as a label for the detection of eicosapantaenoic acid and other fatty acids in serum. These compounds were determined using a reversed-phase column along with a postcolumn reagent containing hydrogen peroxide and a microperoxidase catalyst. An example of a chromatogram obtained with this system is shown in Figure 3.9. The limit of detection for eicosapentaenoic acid was 200 femtomoles.[28] An alternative label, 4-isocyanatophthalhydrazide, has been described by Spurlin and Cooper for use in the detection of derivatized amino acids. In a reversed-phase separation of twelve amino acids, an average limit of detection of 10 femtomoles per amino acid was possible using this label.[29]

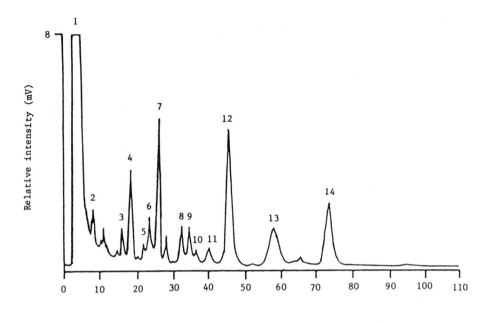

Figure 3.9. Determination of fatty acids in human serum using HPLC and luminol chemiluminescence. Peak identification: 1, ABEI; 2, lauric acid; 3, myristic acid; 4, linolenic acid; 5, eicosapentaenoic acid; 6, palmitoleic acid; 7, unknown substance; 8, linoleic acid; 9, arachidonic acid; 10, docosahexaenoic acid; 11, di-homo-γ-linoleic acid; 12, palmitic acid; 13, oleic acid; 14, margaric acid. (Reproduced with permission from Ref. 28.)

Luminol chemiluminescence can also be used to detect the presence of agents that catalyze the oxidation of luminol. This is done by combining the column eluent with a postcolumn reagent containing an alkaline solution of luminol and hydrogen peroxide. This approach was used by Neary et al. for the detection of Co^{2+} and Cu^{2+} in ion-exchange chromatography.[30] A similar approach has been developed for the detection of heme-containing proteins. The latter was demonstrated by Maltsev and co-workers in the analysis of myoglobin in serum. A size exclusion column was used for the separation of serum proteins, followed by the addition of column eluent to an alkaline luminol/hydrogen peroxide postcolumn reagent.[31]

Because metal ions catalyze luminol chemiluminescence, any analyte that complexes these metal ions (i.e., lowers the metal ion's free concentration) will inhibit light production from luminol. This effect can be used for the indirect detection of proteins and amino acids, both of which complex various metal ions. Based on this approach, Hara and associates have used Cu^{2+}-catalyzed luminol chemiluminescence for the detection of such proteins as thyroglobulin, γ-globulin, albumin, and ovalbumin. Both size exclusion and immunoaffinity columns have been used for this work.[32,33] A similar technique based on Co^{2+}-catalyzed luminol chem-

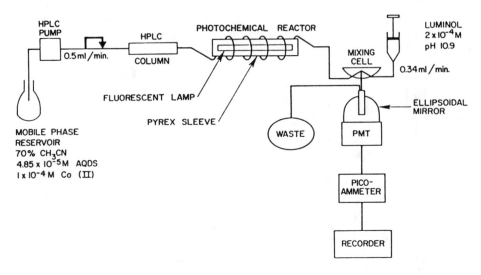

Figure 3.10. System for photochemical production of hydrogen peroxide from analytes for chemiluminescent detection. (Reproduced with permission from Ref. 36.)

iluminescence was used by MacDonald and Nieman for the detection of amino acids. The limits of detection for 12 amino acids varied from 0.004 to 20 nanomoles for 20 µL injections.[34]

As in the peroxyoxalate system, luminol can be used to detect compounds that undergo reactions causing the formation of hydrogen peroxide. The enzymatic generation of hydrogen peroxide has been used by Koerner and Nieman to detect β-D-glucosides separated by reversed-phase liquid chromatography. In their work, enzyme reactors containing immobilized β-glucosidase and glucose oxidase were used along with peroxidase-catalyzed luminol chemiluminescence. The limit of detection was 2 picomoles for a 20 µL injection.[35]

Nonenzymatic reactions can also be used to generate hydrogen peroxide for luminol chemiluminescence. For example, photochemical oxidation was used by Gandelman and Birks to generate hydrogen peroxide from a number of oxygen-containing compounds for chemiluminescent detection. Analytes measured by this approach included aldehydes, aliphatic alcohols, ethers, and sugars. The system used in this work is shown in Figure 3.10. Analytes eluting from the analytical column were first passed through a photochemical reactor in the presence of a photosensitizing agent, anthraquinone-2,6-disulfonate (AQDS). The hydrogen peroxide generated by this reaction was then detected using a postcolumn reagent containing Co^{2+} and luminol.[36]

Oxidants related to hydrogen peroxide can be determined directly using luminol chemiluminescence. For example, Yamamoto et al. demonstrated that various organic peroxides gave sensitivities similar to hydrogen peroxide in microperox-idase-catalyzed luminol chemiluminescence. This was used to detect lipid hydroperoxides in plasma.[37] Miyazawa similarly used cytochrome c–catalyzed luminol chemiluminescence to detect phosphatidylcholine hydroperoxides as they eluted from a silica column. The limit of detection reported was 10 picomoles for a 20 μL injection.[38]

3.4.4. Lucigenin

Lucigenin (N,N'-dimethyl-9,9'-biacridinium dinitrate) is another example of a reagent that has been used for chemiluminescent detection in HPLC. This compound produces light when placed in an alkaline solution containing an oxidant, such as hydrogen peroxide, or an organic reducing agent. The reaction involved in the production of light through lucigenin oxidation is shown in Figure 3.11. This oxidation process results in the formation of an excited molecule of N-methyl acridone, which relaxes to its ground state by emitting light with a wavelength maximum of approximately 470 nm.[3] The quantum yield of this reaction is about 2–3%.[2] The mechanism behind the reaction of lucigenin with organic reducing agents is not well understood. However, this phenomenon is useful since it allows lucigenin to be used in the detection of either oxidizing or reducing substances.[3]

One application of lucigenin chemiluminescence in HPLC has been to detect the reducing agents ascorbic acid and dehydroascorbic acid.[39,40] Other reducing substances detected by lucigenin have included glucose, fructose, lactose, galactose,

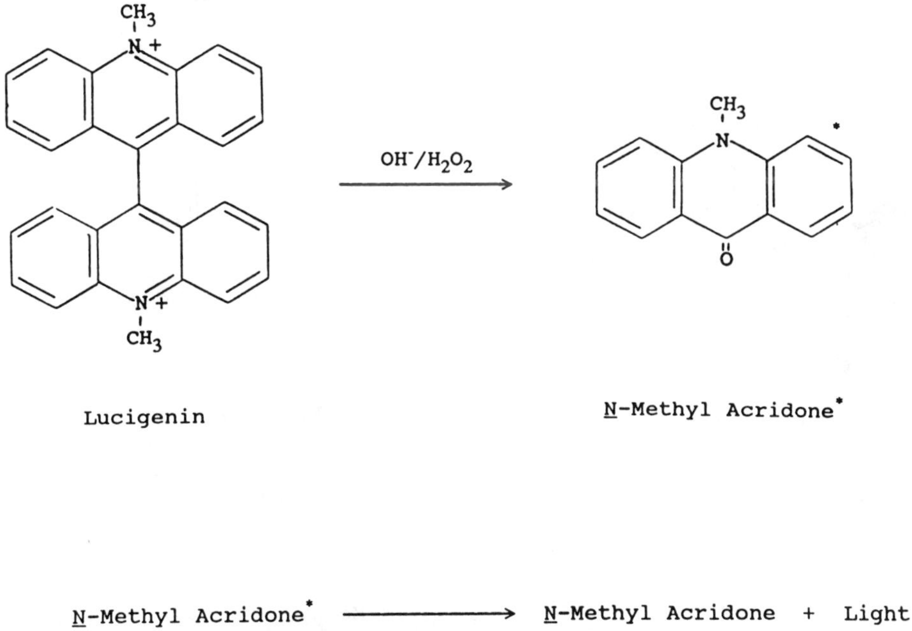

Figure 3.11. Lucigenin chemiluminescence.

and mannose.[40] This has also been used by Maeda and Tsuji in the detection of corticosteroids and p-nitrophenacyl derivatives of carboxylic acids. A separation of six corticosteroids using this approach gave limits of detection of 2 picomoles per injection. A similar separation of seven carboxylic acid derivatives gave limits of detection in the range of 0.5 picomoles per injection.[41]

3.4.5. Acridinium Ester

A chemiluminescent agent closely related to lucigenin is acridinium ester, which is useful as a chemiluminescent label since it emits an intense flash of light in the presence of an alkaline hydrogen peroxide solution. The reactions involved in its light production are shown in Figure 3.12. At high pH, hydrogen peroxide dissociates to form its conjugate base, hydroperoxyl anion (HO_2^-). The hydroperoxyl anion then attacks the acridinium ring, causing a concerted cleavage of bonds and the formation of an excited molecule of N-methylacridone. As in lucigenin chemiluminescence, the excited N-methylacridone then relaxes to its ground state through the emission of light. This light has a maximum emission wavelength at 470 nm and an overall efficiency of approximately 10%.[1,3,42]

One application of acridinium ester labels in HPLC has been their use in the automation of a sandwich immunoassay. This was reported by Hage and Kao, who used acridinium ester–labeled antibodies along with high-performance immunoaffinity chromatography to automate an immunoassay for the detection of parathyroid hormone (PTH) in plasma. The scheme used in automating the immunoassay is shown in Figure 3.13. First, sample and acridinium ester–labeled

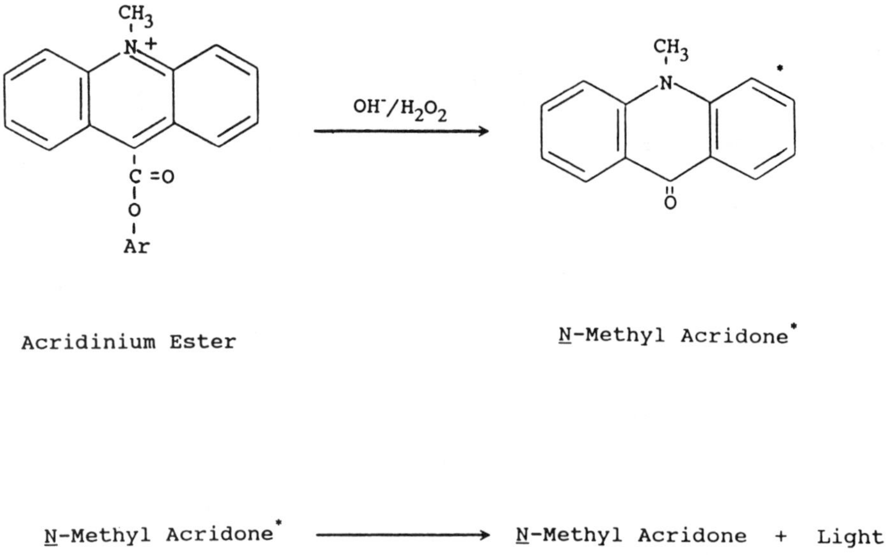

Figure 3.12.   Acridinium ester chemiluminescence.

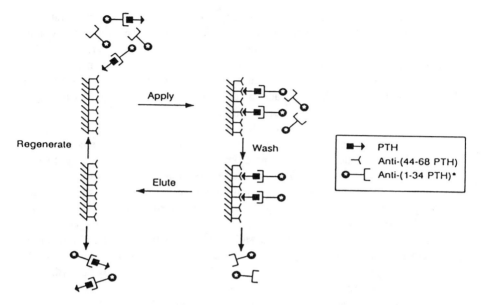

Figure 3.13. Automated sandwich immunoassay for parathyroid hormone (PTH) using high-performance immunoaffinity chromatography and acridinium ester chemiluminescence. The asterisk (*) indicates the acridinium ester label. (Reproduced with permission from Ref. 43.)

anti-(1-34 PTH) antibodies were injected onto a column containing immobilized anti-(44-68 PTH) antibodies. After washing excess label and nonretained components from the column, the retained PTH and associated labeled antibody were eluted and detected by combining them with an alkaline hydrogen peroxide postcolumn reagent. The limit of detection for PTH was 16 attomoles for a 66 μL injection, and the total chromatographic time was only 6 min per injection, following a 1 h incubation of sample and labeled antibody. By using different immunoaffinity columns and labeled antibodies, this technique could be adapted easily for the detection of other large biological compounds.[43]

3.4.6. Other Chemiluminescent Systems

In addition to the common systems already discussed, a number of other reactions have also been used for chemiluminescent detection in HPLC. For example, Abbott et al. developed a method for the determination of morphine based on the native chemiluminescence of this drug when placed in a solution containing permanganate and acid tetraphosphate. This technique was used to detect as little as 20 picomoles of morphine in extracted blood or urine samples. The extension of this method to other drugs that display native chemiluminescence should be possible.[44] A method for the detection of proteins based on the Cu^{2+}-catalyzed chemiluminescent reaction between 1,10-phenanthroline and hydrogen peroxide has been demonstrated by Hara et al. As in the case of the luminol reaction, proteins were detected by their ability to complex Cu^{2+}, resulting in a decrease in light production by 1,10-phenanthroline. This method has been used for detection in both metal chelate affinity chromatography[45] and immunoaffinity chromatography.[46] Proteins that have been measured by this approach include bovine serum albumin, bovine serum γ-globulin, lysozyme,[45] and human serum albumin.[46]

Malcolme-Lawes and co-workers have examined the use of electrochemiluminescence in the detection of various organic compounds. This approach is based on the ability of some aromatic compounds to produce light when placed between two electrodes in a conducting solution. Various aspects have been examined in the design of electrochemiluminescence detection systems for reversed-phase liquid chromatography. Experimental aspects considered include the use of dc versus low-frequency ac current, [47,48] the design of the flow cell,[49] and the use of derivatization in analyte detection.[50] A variety of standard compounds have been detected with this technique. Examples include naphthalene, pyrene, carbazole, chrysene,[47,48] various phthalates,[49] and DNP- or phenylthiohydantoin derivatives of amino acids.[50]

A number of investigators have examined the use of HPLC detectors based on chemiluminescence induced by ozone or singlet oxygen. In work by Birks and co-workers, a system was developed in which column eluent was converted into an aerosol spray using a high-velocity stream of ozone and oxygen. In the presence of this gas, certain analytes (e.g., olefins, divalent sulfur compounds, hydrazines, azides, nitrogen heterocyclics, and highly fluorescent compounds) are excited and emit light. Detection limits vary with the analyte, but are generally in the low microgram to picogram range.[51,52]

Ozone-induced chemiluminescence has also been used to detect compounds that can either be pyrolyzed or chemically treated to produce nitrogen oxide radicals. The nitrogen oxide is then reacted with ozone to produce an excited molecule of nitrogen dioxide. The nitrogen dioxide later relaxes to its ground state, resulting in light production. In work by Yu and Goff, the nitrogen oxide/ozone reaction was used to quantify nitrate esters of glycerol, isosorbide, and pentaerythritol. Nitrogen oxide was produced by pyrolyzing these compounds after they were separated on a normal-phase column. Limits of detection ranged from 0.1 to .2 nanograms for a 25 μL injection.[53] DeAngelis et al. developed a method for the determination of organic molecules in which nitrogen oxide was produced by the oxidation of these compounds with dilute nitric acid, nitrous acid, or nitrite ions.[54] Conboy and Hotchkiss used chemiluminescence and a photolytic reactor to monitor nitrogen oxide produced by N-nitrosoamino acids and N-nitrosamides.[55] McNamara et al. used a superconducting metal oxide catalyst to form nitrogen oxide from organic compounds by their reaction with nitrogen dioxide. The method was capable of detecting alcohols, acetaldehyde, acetone, methylethylketone, acetonitrile, ammonia, and nitromethane.[56] The nitrogen oxide/ozone reaction has also been used in the detection of nitrated polycyclic aromatic hydrocarbons by Robbat and co-workers.[57]

3.5. Conclusion

In summary, chemiluminescence is of growing interest as a way of obtaining highly sensitive and selective detection in HPLC. The apparatus for these measurements is relatively simple and yet allows extremely low quantities of analyte to be measured. Also, a number of chemiluminescent reactions are available for use in HPLC. These include methods based on bioluminescence, oxalate esters, luminol, lucigenin, and acridinium ester. Several other systems, such as those based on ozone-induced chemiluminescence and electrochemiluminescence, have also been developed for HPLC detection. These chemiluminescent systems can be used to detect a wide variety of compounds, including amino acids, carboxylic acids, proteins and peptides, steroids, aromatics, alcohols, sugars, and others. In many cases, picomole or femtomole quantities of analyte can be determined. Because of these many attractive characteristics, it is expected that chemiluminescence will see increased use as a means for detection in HPLC.

3.6. References

1. A. K. Campbell, *Chemiluminescence,* VCH Publishers: New York (1988).

2. G. J. de Jong and P. J. M. Kwakman, *J. Chromatogr.* **492** (1989) 319–43.

3. M. L. Grayeski, *Anal. Chem.* **59** (1987) 1243A–1256A.

4. W. R. Seitz and M. P. Neary, in *Contemporary Topics in Analytical Chemistry,* D. M. Hercules, Ed., Plenum: New York, (1977), Vol. I, 49.

5. W. P. Bostick, M. S. Denton, and S. R. Dinsmore, in *Bioluminescence and Chemiluminescence, Instruments and Applications,* K. Van Dyke, Ed., CRC: Boca Raton, FL (1985), Vol. II, 227–46.

6. K. Arisue, Y. Marui, T. Yoshida, Z. Ogawa, K. Kohda, C. Hayashi, and Y. Ishidi, *Rinsho Byori* **29**(5) (1981) 459–62.

7. K. Imai, *Methods Enzymol.* **133** (1986) 435–49.

8. M. M. Rauhut, *Acc. Chem. Res.* **2** (1969) 80–87.

9. K. W. Sigvardson and J. W. Birks, *Anal. Chem.* **55** (1983) 432–35.

10. M. L. Grayeski and A. J. Weber, *Anal. Lett.* **17** (1984) 1539–52.

11. K. W. Sigvardson, J. M. Kennish, and J. W. Birks, *Anal. Chem.* **56** (1984) 1096–102.

12. K. W. Sigvardson and J. W. Birks, *J. Chromatogr.* **316** (1984) 507–18.

13. M. L. Grayeski and J. K. DeVasto, *Anal. Chem.* **59** (1987) 1203–6.

14. G. Mellbin, *J. Liq. Chromatogr.* **6** (1983) 1603–16.

15. S.-I. Kobayashi, J. Sekino, K. Honda, and K. Imai, *Anal. Biochem.* **112** (1981) 99–104.

16. O. Nozaki and Y. Ohba, *Anal. Chim. Acta* **205** (1988) 255–60.

17. T. Koziol, M. L. Grayeski, and R. Weinberger, *J. Chromatogr.* **317** (1984) 355–66.

18. P. J. M. Kwakman, U. A. Th. Brinkman, R. W. Frei, G. J. de Jong, F. J. Spruit, N. G. F. M. Lammers, and J. H. M. van den Berg, *Chromatographia* **24** (1987) 395–99.

19. G. J. de Jong, N. Lammers, F. J. Spruit, U. A. Th. Brinkman, and R. W. Frei, *Chromatographia* **18** (1984) 129–33.

20. K. Miyaguchi, K. Honda, T. Toyo'oka, and K. Imai, *J. Chromatogr.* **352** (1986) 255–60.

21. K. Honda, K. Miyaguchi, and K. Imai, *Anal. Chim. Acta.* **177** (1985) 111–20.

22. K. Honda, K. Miyaguchi, H. Nishino, H. Tanaka, T. Yao, and K. Imai, *Anal. Biochem.* **153** (1986) 50–53.

23. P. van Zoonen, C. Gooijer, N. H. Velthorst, and R. W. Frei, *J. Pharm. Biomed. Anal.* **5** (1987) 485–92.

24. P. van Zoonen, I. de Herder, C. Gooijer, N. H. Velthorst, and R. W. Frei, *Anal. Lett.* **19** (1986) 1949–61.

25. H. Jansen, U. A. Th. Brinkman, and R. W. Frei, *J. Chromatogr.* **440** (1988) 217–23.

26. A. C. Capomacchia, R. N. Jennings, S. M. Hemingway, P. D'Souza, W. Prapaitrakul, and A. Gingle, *Anal. Chim. Acta* **196** (1987) 305–10.

27. T. Kawasaki, M. Maeda, and A. Tsuji, *J. Chromatogr.* **328** (1985) 121–26.

28. H. Yuki, Y. Azuma, N. Maeda, and H. Kawasaki, *Chem. Pharm. Bull.* **36** (1988) 1905–8.

29. S. R. Spurlin and M. M. Cooper, *Anal. Lett.* **19** (1986) 2277–83.

30. M. P. Neary, R. Seitz, and D. M. Hercules, *Anal. Lett.* **7** (1974) 583–90.

31. V. G. Maltsev, T. M. Zimina, A. B. Khvatov, and B. G. Belenkii, *J. Chromatogr.* **416** (1987) 45–52.

32. T. Hara, M. Toriyama, and T. Ebuchi, *Bull. Chem. Soc. Jpn.* **58** (1985) 109–14.

33. T. Hara, M. Toriyama, T. Ebuchi, and M. Imaki, *Bull. Chem. Soc. Jpn.* **59** (1986) 2368–70.

34. A. MacDonald and T. A. Nieman, *Anal. Chem.* **57** (1985) 936–40.

35. P. J. Koerner, Jr., and T. A. Nieman, *J. Chromatogr.* **449** (1988) 217–28.

36. M. S. Gandelman and J. W. Birks, *J. Chromatogr.* **242** (1982) 21–31.

37. Y. Yamamoto, M. H. Brodsky, J. C. Baker, and B. N. Ames, *Anal. Biochem.* **160** (1987) 7–13.

38. T. Miyazawa, K. Yasuda, K. Fujimoto, and T. Kaneda, *Anal. Lett.* **21** (1988) 1033–44.

39. R. L. Veazey and T. A. Nieman, *J. Chromatogr.* **200** (1980) 153–62.

40. R. L. Veazey, H. Nekimken, and T. A. Nieman, *Talanta* **31** (1984) 603–6.

41. M. Maeda and A. Tsuji, *J. Chromatogr.* **352** (1986) 213–20.

42. I. Weeks, M. Sturgess, R. C. Brown, and J. S. Woodhead, *Methods Enzymol.* **133** (1986) 366–87.

43. D. S. Hage and P. C. Kao, *Anal. Chem.* **63** (1991) 586–95.

44. R. W. Abbott, A. Townshend, and R. Gill, *Analyst* **112** (1987) 397–406.

45. T. Hara, K. Tsukagoshi, and T. Yoshida, *Bull. Chem. Soc. Jpn.* **61** (1988) 2779–83.

46. T. Hara, K. Tsukagoshi, A. Arai, and T. Iharada, *Bull. Chem. Soc. Jpn.* **61** (1988) 301–3.

47. C. Blatchford and D. J. Malcolme-Lawes, *J. Chromatogr.* **321** (1985) 227–34.

48. C. Blatchford, E. Humphreys, and D. J. Malcolme-Lawes, *J. Chromatogr.* **329** (1985) 281–84.

49. E. Hill, E. Humphreys, and D. J. Malcolme-Lawes, *J. Chromatogr.* **370** (1986) 427–37.

50. E. Hill, E. Humphreys, and D. J. Malcolme-Lawes, *J. Chromatogr.* **441** (1988) 394–99.

51. J. W. Birks and M. C. Kuge, *Anal. Chem.* **52** (1980) 897–901.

52. B. Shoemaker and J. W. Birks, *J. Chromatogr.* **209** (1981) 251–63.

53. W. C. Yu and E. U. Goff, *Anal. Chem.* **55** (1983) 29–32.

54. J. J. DeAngelis, R. M. Barkley, and R. E. Sievers, *J. Chromatogr.* **441** (1988) 125–34.

55. J. J. Conboy and J. H. Hotchkiss, *Analyst* **114** (1989) 155–59.

56. E. A. McNamara, S. A. Montzka, R. M. Barkley, and R. E. Sievers, *J. Chromatogr.* **452** (1988) 75–83.

57. A. Robbat, Jr., N. P. Corso, and T.-Y. Liu, *Anal. Chem.* **60** (1988) 173–74.

CHAPTER

4

HPLC Detection in the Near Infrared

Gabor Patonay and Tibor Czuppon

Department of Chemistry
Georgia State University
Atlanta, Georgia 30303

4.1. Introduction

The problems posed by new biotechnologies has led to the development of new analytical techniques that are more specific within a complex biological system. Recently a new approach to absorption or fluorescence detection in HPLC was developed using the longer-wavelength portion of the electromagnetic spectrum. This chapter is not designed to provide an exhaustive discussion of the different methods that may be used to achieve more sensitive and selective detection of chromatography eluents by the utilization of the longer-wavelength part of the electromagnetic spectrum. Rather, this chapter deals with the topic in an introductory fashion to illustrate the number of possibilities that are open for the scientist. Here we seek only a representative, as opposed to an exhaustive, survey of the HPLC application and methodology of near-infrared fluorescence. Due to the nature of this relatively young approach, some of our discussions will be necessarily speculative. The main goal of this chapter is to provoke the interest of the scientific community in this new and powerful spectroscopic HPLC detection method.

Absorption or fluorescence spectroscopy has long been an important detection method in high-performance liquid chromatography (HPLC) separations.[1-3] The detection mechanism is based on the observation of electronic transitions [4,5] in molecules. Because these transitions usually require higher-energy photons, they may be observed in the ultraviolet or visible part of the electromagnetic spectrum. However, there are several molecules whose electronic energy levels are closely spaced, and lower energy is required for these electronic transitions. The following chapter reviews spectroscopic HPLC detection methods that involve electronic transitions in the longer-wavelength part of the electromagnetic spectrum, that is, in the near-infrared (NIR) part of the spectrum.

Because of its selectivity and sensitivity, fluorescence has proved to be a valuable detection method in HPLC analyses. Fluorescence detection methods have been extensively applied in different separation methods, including a wide array of bioanalytical separations.[6-8] However, this latter application may be severely limited if the interference from the biological matrix is large. The interference is most significant in the UV or the short-wavelength visible region. This explains the continuous effort of the scientific community to find fluorescence detection methods of lesser interference. One possibility for countering this problem is the extension of fluorescence detection methods into the longer-wavelength region of the electromagnetic spectrum.[9,10] It is important to realize, however, that most of the solutes that are typically separated on an HPLC column do not exhibit significant fluorescence in this longer-wavelength region. Accordingly, additional chemical, physical, and analytical methods are required to induce significant fluorescence in this spectral region. These methods are similar to the methods used for inducing fluorescence in the shorter-wavelength region, for example, derivatization, direct or indirect fluorescence labeling, etc.[11] Often, however, due to the differences in the chemistry of the NIR absorbing chromophores, the techniques used in the visible region may not be directly applicable.

The near-infrared spectral region is one of the lowest in interference. Very few fluorophores exhibit native fluorescence in this spectral region, which extends up to 1100 nm.[12] Unfortunately, derivatization methods have limited applications in this region. The only viable method that can be used to increase the NIR fluorescence properties of the solute is direct or indirect labeling using NIR absorbing fluorophores. When the chromatographer combines semiconductor laser excitation with the application of NIR fluorophore labeling, an additional advantage is gained.[13] In most cases, since only the NIR fluorophore exhibits significant fluorescence, the interference is reduced. Simultaneously, laser diode technology increases the sensitivity of the technique, without additional financial burden. Although NIR laser diode HPLC detection techniques are relatively new, the range of applications discussed in this chapter points to the potential of NIR HPLC detection methods in the analysis of chemical and biochemical separations. The first part of this chapter presents an introduction to the theory and methods of NIR HPLC detection methods. Following this theoretical background, several bioanalytical applications are discussed. Finally, examples of NIR laser diode excitation in HPLC detections will be presented to demonstrate the broad applicability and selectivity of NIR HPLC detection methods.

To fully understand the advantages of using NIR fluorescence spectroscopy in the detection of HPLC eluents, it is necessary to look at some of the special features of NIR fluorophores as well as the special problems associated with laser diode excitation. It is also important to discuss the chemistry related to the development of NIR chromophores that may be used to enhance the luminescence response of the HPLC solute of interest in the longer-wavelength part of the electromagnetic spectrum.

4.2. Fluorescence Properties of NIR Absorbing Chromophores

Fluorescence spectroscopy has been applied to detection of HPLC eluents for several decades. Fluorescence HPLC detection has been described in several excellent monographs, and therefore we will only discuss some special aspects of NIR fluorescence that are important in the application for HPLC detection.

Almost invariably, NIR fluorophores undergo an excitation process resulting in the population of the first singlet excited state.[14-16] As is the case with their visible counterparts, the molar absorptivity of the chromophore determines the probability of absorption of electromagnetic radiation. Fortunately, most NIR fluorophores are advantageous in this respect; that is, their typical molar absorptivity values are larger than 100,000. For some NIR fluorophores it can approach 300,000. Consequently, the probability of photon absorption is high, making low detection limits possible. The efficiency of NIR fluorescence is also very important concerning low detection limits in HPLC separations. The fluorescence quantum yield, which is defined as the ratio of the emitted number of photons relative to the number of photons absorbed to the singlet excited state, plays an important role in achieving low detection limits. Due to the relatively high probability of different competing processes, the fluorescence quantum yield of NIR chromophores may be relatively low, especially if the molecule has less rigid structures typical for heptacyanines.

There are two major families of NIR absorbing chromophores that can be used as NIR fluorescent tagging molecules, the cyanine and the phthalocyanine dye families.[12,14,17] The major requirements for any NIR absorbing chromophore used in HPLC determinations are: high fluorescence quantum yield, large Stokes shift, availability of excitation maximum for laser diode excitation, chemical and photochemical stability, low susceptibility to fluorescence quenchers that may be present in the matrix, and the presence of appropriate reactive functional groups that may be used for coupling of the NIR reporting group to the analyte molecule of interest. The relatively long excitation and fluorescence wavelengths require the presence of extensive conjugation in the molecule. In both the cyanine and the phthalocyanine dye families, we can observe this extensive conjugation. In addition to these two groups of molecules, there are certain proteins, the phycobiliproteins, that can exhibit significant fluorescence in the NIR region. Due to their large molecule weight, however, their analytical application is limited. A huge number of dyes are known in both the cyanine and the phthalocyanine families,[12,14,17] and several of these compounds have good fluorescence quantum yield, making them prime candidates for use in NIR HPLC detection. Since the fluorescence and absorbance maxima of these dyes are dependent on the chemical structure of the fluorophore, the researcher can choose spectroscopic properties that match the desired excitation wavelength, as discussed in a latter part of this chapter.

There are basically two ways we can utilize NIR fluorogenic labels for HPLC detection depending on the chemistry we wish to follow during the labeling process.

Depending on the chemical sensitivity of the analyte of interest, the researcher may choose covalent or noncovalent labeling methods.[13,18] If the covalent method is chosen, the analyte molecule and the NIR label molecule should have appropriate functional groups. If noncovalent labeling is chosen, then basically nonspecific binding of the NIR chromophore will be used through adsorption, electrostatic forces, or other similar processes. This latter approach should provide suitable stability during the chromatographic process. At this early stage of the development of this new analytical method, due to the limited commercial availability of reagents, the noncovalent labeling method is preferred.

4.3. Laser Diodes

In order to understand the difficulties associated with the use of semiconductor laser sources in HPLC separations, we need to review some special characteristics of laser diodes. Laser diodes are special semiconductor devices that are able to emit monochromatic, coherent laser radiation. Since most of the laser diodes are similar in structure to light-emitting diodes, it is useful to treat both similarly in our discussions. As a matter of fact, modern high-power-output light-emitting diodes (LEDs) can be equally useful in HPLC detector applications. The most important difference between the LEDs and laser diodes regarding HPLC detection techniques is the monochromaticity of the emitted radiation. LEDs have a 10–30 nm bandwidth, making their output much less monochromatic. Monochromaticity, however, is not necessarily a requirement if NIR absorbing chromophores are used as reporting labels. Most of the NIR fluorophores exhibit a relatively wide absorption peak in the NIR spectral region. A typical peak width of several tens of nanometers for LEDs may not be a serious disadvantage if the reporting group has a wide absorption peak, as with cyanine dyes. When phthalocyanines or naphthalocyanines are used as reporting groups, the match between the fluorophore and light source is more important, since these latter fluorophores have much narrower absorption characteristics. To aggravate the problem further, naphthalocyanines usually have a very small Stokes shift, which creates additional difficulties.

Both laser diodes and LEDs are constructed using semiconductor material. The forward-biased current flow with the recombination of electrons and holes in the semiconductor material releases energy at the junction in the form of a photon with energy roughly equivalent to the band gap between the valence and conduction band. Light emission occurs spontaneously, at random, in LEDs, as opposed to laser diodes, where stimulated emission dominates. In most applications, however, we need to be aware that a certain percentage, albeit a small one, of the emitted radiation of the diode lasers is random, not stimulated radiation. This portion of the output radiation of laser diodes is very much like the radiation from LEDs. Since this part of the radiation is non-coherent and non-monochromatic, it needs to be minimized. Several types of applications require that filters be present to remove the nonlaser radiation from the beam. It is important to realize, however, that some

Wavelength vs. Temperature

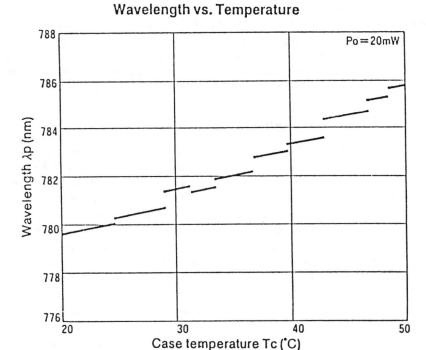

Figure 4.1. Typical output characteristics of a semiconductor laser diode (Sharp LT024MFO) as a function of the ambient temperature (by permission). by Sharp

HPLC applications may not require the use of laser diode excitation sources; high-intensity LEDs may be used with equal efficiency.

The overall efficiency of laser diodes is determined by the threshold current and slope efficiency. Typical NIR laser diode efficiency may reach 10% or higher. The spectral bandwidth of laser diodes is typically a few tenths of a nanometer. The output wavelength of laser diodes depends on the temperature of the semiconductor junction, which allows one to change the output wavelength by changing the temperature of the laser diode. This temperature tuning, however, does not result in a continuous changing of the output wavelength, because laser diodes can oscillate simultaneously on different modes. As the temperature of the laser diode changes, the laser may mode hop from one mode to another, abruptly changing the output wavelength (Figure 4.1). To avoid wavelength instabilities, the laser diodes need to be kept at constant temperatures, usually by using Peltier elements.

Without corrective optical elements, the beam quality for laser diodes is very poor. This is due to the short cavity and the fact that the emitting area is much wider than higher. Generally laser diodes produce rapidly diverging, oval-shaped beams with relatively poor polarization. The emitting area of a single-mode laser is only few micrometers long, causing the emitted beam to spread out at a 40° angle perpendicular to the junction and about a 10° angle in the plane of the junction. This low-quality beam is, however, not significant in most HPLC applications. With appropriate focusing elements the beam can be sufficiently focused for use with small detector volumes. It is possible to focus the laser diode beam into a spot about 50 μm or less in diameter, which is sufficient for certain capillary applications.

4.3.1. Intracavity Applications

One possible use of laser diodes in HPLC detectors is measuring absorbance in the NIR region. It can be achieved with the conventional arrangement, where the laser diode is only used as a light source. A separate detector is required for determining the analyte absorbance in this case. A more cost-effective application is when the laser diode is operated in the intracavity mode, as described by Patonay and co-workers.[19,20] In this case no additional detector is required, since the builtin photodiode of the laser diode is used as a detector to determine analyte absorbencies. Since the intracavity application of laser diodes has been discussed earlier, only special problems associated with HPLC applications will be reviewed in this chapter.

Commercial laser diodes are available in low-, medium-, and high-power versions. The high-power versions are usually multistripe devices and therefore less suitable for intracavity applications, since these lasers typically have as many as 5 to 40 cavity stripes. The low- and medium-power versions are usually single-stripe devices; for example, there is only one active lasting cavity. The intracavity application is further complicated by the differences in the reflectivity of the facets of the laser diode chip. The commercially available low-power laser diodes have a medium reflectivity coating, typically $R = 0.3$, at both facets. These diodes are available for several consumer applications, such as CD players, bar code readers, etc.; so the effect of optical feedback is intentionally minimized. Consequently, they are less suitable for intracavity applications. The medium-power versions (up to 30 mW) have a different chip structure. In order to achieve higher optical output, the back-reflecting surface is coated with a highly reflective multilayer film composed of amorphous silicon and aluminum oxide, and the front facet is coated with a transparent dielectric film of aluminum oxide. This latter has a double role: It acts as a passivation coating and provides the required reflectivity. This chip structure efficiently directs the laser beam forward out of the chip. These medium-powered laser diodes are more suitable for intracavity applications due to their highly transparent front facet surface.

Laser diodes are also available with antireflection-coated front facets. They have an additional layer of zirconium dioxide over the front facet for a total optical quarter-wave thickness. The front facet reflectivity is so low in these devices that it is very difficult to measure. Due to the very low front facet reflectivity, the anti-reflection-coated diodes are the most suitable devices for intracavity applications. The major advantage of using laser diodes as intracavity detectors arises from the fact that the increased cavity loss due to analyte absorption, may be compensated for by increasing the forward driving current.

The most important aspect of HPLC applications of laser diode intracavity spectrometry is that the beam needs to be collimated so that the beam diameter is less than the width of the flow cell. Due to the operation of the laser diode intracavity spectrophotometer, only those flow cells that have a square channel inside the flow cell are usable. This arrangement would minimize refraction of the beam. Since the laser beam makes more than one pass in intracavity operation, minimizing unwanted refractance is essential. Accordingly, the use of capillary arrangements is not beneficial. Due to the refraction of the curved surfaces of the capillary column, intracavity operation is not possible.

4.4. Near-Infrared Fluorophore Labels

As discussed earlier, most analyte molecules of interest do not absorb in the NIR spectral region. Hence, it is essential to introduce this property into the solute molecule, so that it can be detected in the NIR spectral region. There are several different possibilities to introduce NIR absorbing chromophores into the analyte molecule of interest. These methods may be characterized as physico-chemical or chemical methods depending on the nature of the bond between the label and the analyte molecule. Often a simple physico-chemical interaction between the analyte and the reporting label may be suitable for labeling purposes. Under this category, for example, ion-pair formation may be used if both the analyte and the dye molecule are charged. Polymethine cyanine dyes, which are ionic molecules, may be used for labeling ionic analytes, such as surfactants.[21] Ion pairing, however, in most cases, is not adequate for labeling use in HPLC separations.

Competitive binding of NIR fluorophores has been utilized for labeling proteins during HPLC separations. Kamisaka et al. studied[18] the competitive binding of indocyanine green to human and bovine serum albumin for HPLC detection purposes. ICG was chosen as the label because it is known to bind strongly to serum albumins. In spite of this strong interaction, however, the complex tends to dissociate on the chromatography column, making the procedure less effective. It was observed that the absorption and fluorescence spectra of the ICG change significantly upon binding to serum albumins. This change is a result of the lower dimerization of the fluorophore that occurs in aqueous solutions. Additional studies indicated that other NIR absorbing fluorophores may also exhibit significant changes upon complexation to serum albumin. All these studies indicate that the use of non-covalently bound labels in HPLC may require careful evaluation. Indocyanine green itself has been used for several other applications, for example, NIR laser dye, medical diagnostic tool, etc. Although indocyanine green does not have large fluorescence quantum efficiency, its excellent binding characteristics make it useful in HPLC separations. The chemical structure of indocyanine green is depicted in Figure 4.2.

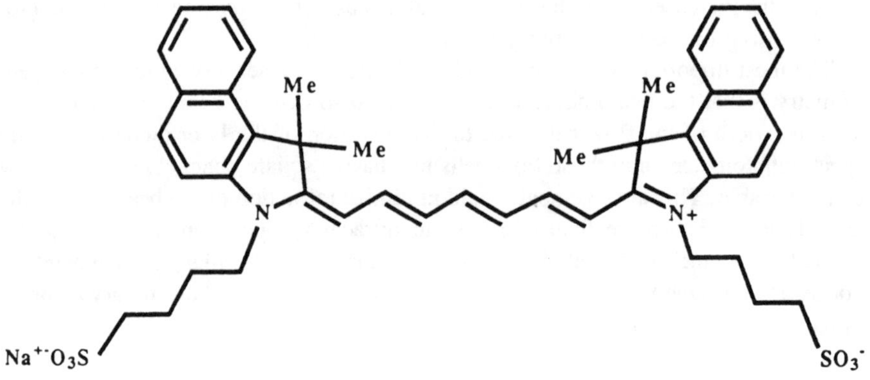

Figure 4.2.   Chemical structure of indocyanine green (ICG).

The binding characteristics of indocyanine green have been studied extensively. These studies served as an important foundation for the application of indocyanine green as a noncovalent label in HPLC separations. The first bioanalytical applications of this dye were reported by Imasaka and Ishibashi using NIR fluorescence. Their work employed human serum albumins with indocyanine green using electrostatic and hydrophobic interactions. The labeled proteins then were separated on a gel filtration column and detected using an NIR fluorescence. The NIR light source used for excitation was NIR-emitting laser diode. Low detection limits and enhanced detection of the serum component were reported in these studies. Using this method, the detection of other proteins is also possible after chromatographic separation. In spite of the low efficiency of this type of labeling, the utility of the method was clearly demonstrated. Picomolar detection limits were reported for NIR-labeled proteins using this detection method in spite of its relatively ineffective detection system. It is possible to increase detection limits significantly by replacing the relatively low-efficiency photomultiplier tube with a more efficient semiconductor detector, such as a silicon photodiode or avalanche photodiode. In Figure 4.3, we can see a typical chromatogram for human serum albumin using NIR fluorescence detection. Note the stable baseline, much less noisy than the baseline typical in the UV region. All these features are characteristic of NIR HPLC detection methods that result from lower background interference. Although problems do exist with the use of a noncovalent dye, the utility of NIR fluorescence for HPLC detection was amply demonstrated in these studies. We need to point out an additional advantage of NIR detection methods. Human serum lipoprotein exhibits an absorptivity so low that conventional UV detection is not able to reveal information about the true relative concentration of this component in human serum. This is amply demonstrated in Figure 4.4.

Williams et al.[22] evaluated the use of covalent and noncovalent NIR-absorbing fluorophores in HPLC separations. Both reverse-phase and size-exclusion techniques were evaluated in conjunction with NIR labeling techniques. The binding

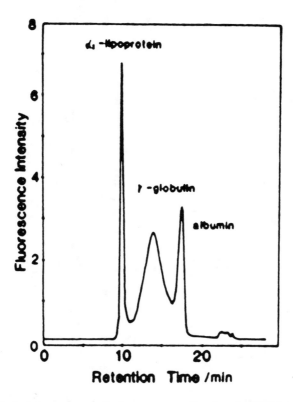

Figure 4.3. Typical chromatogram for human serum albumin using NIR detection (Ref. 13, by permission).

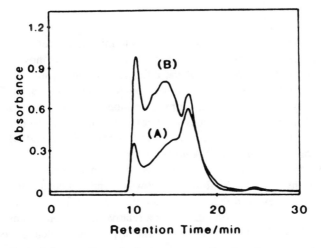

Figure 4.4. Typical chromatogram for human serum albumin using conventional (A) UV detection and (B) NIR labeling (Ref. 13, by permission).

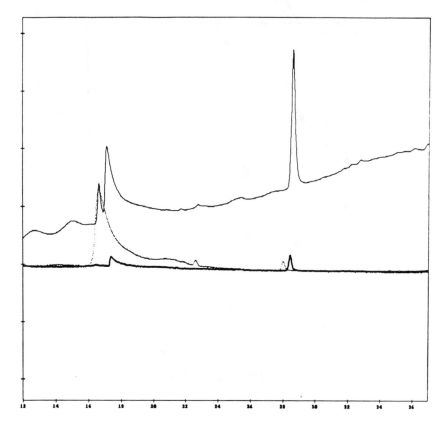

Figure 4.5. Comparison of covalent and noncovalent labeling during HPLC separation of NIR labeled analytes. (UV detection – , NIR detection –)

constant for the noncovalent labeling was determined, indicating that in certain cases noncovalent labeling may be feasible. It was noticed, however, that non-covalent labeling tends to decrease the efficiency of the separation. Much better separation was achieved when covalent labeling techniques were used. Figure 4.5 is a typical representation of chromatograms obtained using covalent labeling.

Additional interesting differences were observed when noncovalent and covalent labeling methods were compared using reversed-phase HPLC and NIR absorption detection. Covalently labeled HSA protein peaks indicated uniform labeling at both hydrophobic and hydrophilic binding sites, while noncovalent labeling showed preference for hydrophobic binding sites. It was found that covalently labeled HSA peaks were better resolved than noncovalently labeled peaks when reversed-phase HPLC was used. The C18 column creates a relatively harsh environment for HSA. It is not surprising that some of the noncovalently labeled protein dissociated form its label.

There are other NIR-absorbing chromophores besides the hydrophobic carbo-cyanine dyes that need to be mentioned. Phycobiliproteins, which are large-mo-lecular-weight compounds, exhibit strong NIR absorbance. These compounds may be directly conjugated to large proteins, biotin, or avidin for labeling purposes. The major advantage of these compounds is the relatively high molar absorbtivity and good quantum yield. The relatively high molecular weight of the label, however, limits its applicability in HPLC separations.

Porphyrins and porphyrinlike compounds also exhibit absorbance and fluores-cence in the NIR spectral region. Chemical and light stability is much more advan-tageous in these molecules. Phthalocyanines have typical absorption wavelengths matching the output wavelengths of the 680 nm laser diodes, while naphthalo-cyanines have typical absorption wavelengths that match the output wavelength of the 780 nm laser diodes. This latter group of dyes is more advantageous because the longer absorption wavelengths mean there is interference from the matrix. Also the output power and efficiency of the laser diodes are much better in this region than in the 680 nm region, adding further advantages. The carbocyanine dyes are ionic compounds that may create problems in certain separations, especially when used to label smaller molecules.

We have investigated the application of aluminum naphthalocyanines for labeling purposes. These chromophores are especially advantageous, since the central metal may be used to incorporate a relatively simple linker group into the NIR chro-mophore. A typical preparation of an NIR-absorbing naphthalocyanine dye is as follows: 0.18 g (1.01 mmol) of 2,3-dicyanonaphthalene, and 1.0 g (16.6 mmol) urea were melted at 180°C for 10 min and then were allowed to cool to room temperature. 0.1 g (0.75 mmol) anhydrous $AlCl_3$ was added to the reaction mix-ture in N_2 atmosphere in a pressure tube. The pressure tube was then sealed and heated for 1 h in a 240°C oil bath. The reaction mixture was then allowed to cool to room temperature and was first boiled with 20 mL 5% HCl, followed by boiling with 5% NaOH. After filtration, the residue was washed with distilled water to pH 7

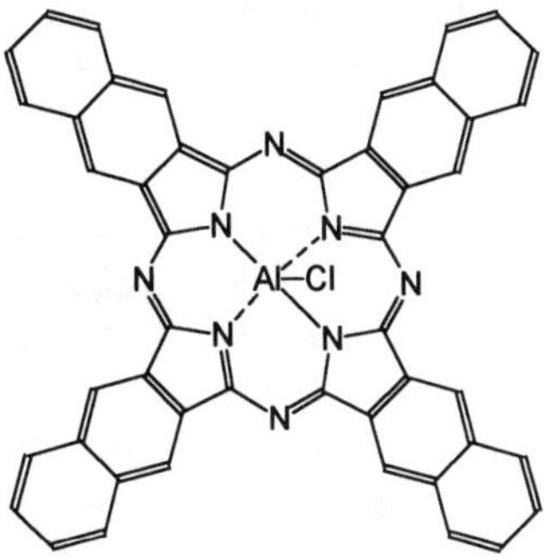

Figure 4.6. Chemical structure of the naphthalocyanine dye used in this study.

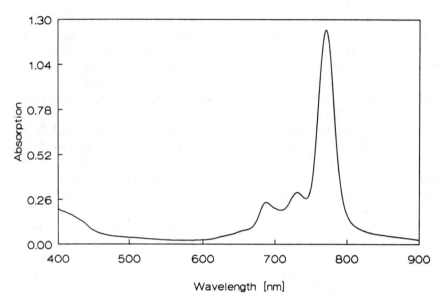

Figure 4.7. Typical absorption curve for the naphthalocyanine dye used in this study.

and dried at 120°C. The chemical structure of the dye is depicted in Figure 4.6. Figure 4.7, a typical absorption curve for this aluminum naphthalocyanine, indicates the absorption maxima of the chromophore in the desired low-interference region. It can be also seen from this figure that the absorption maximum matches perfectly the typical output wavelength of the commercially available and very inexpensive 30 mW laser diode. It should be noted, however, that the absorption band is much narrower than for the carbocyanine dyes, requiring more matched light sources. Naphthalocyanines are extremely stable compounds, although their aqueous solubility is limited. Aqueous solubility may be increased by introducing water-soluble functional groups into the dyes.

The detection limit in the NIR region is determined by factors similar to those in other wavelength regions. Fluorescence detection is preferred when low detection limits are desirable. Sauda et al. reported[13] the use of a laser-diode-based detector for HPLC detection. The detection limit reported was 1.3 pmol for albumin. Better detection limits may be achieved if semiconductor detectors are used in which spectral sensitivity matches the fluorescence wavelength of the label. Figure 4.8 is an example of an HPLC detector design that fulfills these requirements. As can be seen, a semiconductor laser is used for excitation. The beam is focused to illuminate the center of the flow cell. Any regular fluorometric flow cell including capillary detectors may be used in this arrangement. The output of the semiconductor laser is passed trough an interference filter for better spectral purity. The fluorescence signal is measured at the usual 90° angle using semiconductor diodes, preferably silicon or avalanche photodiodes. To obtain the highest throughput by the optics, an inter-

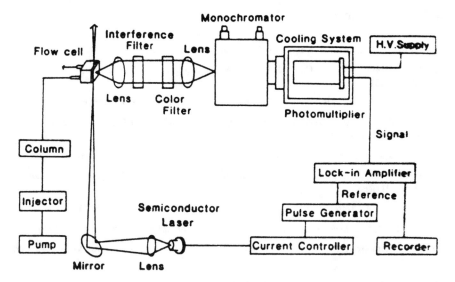

Figure 4.8. Schematic of the HPLC NIR detector used by Sauda et al. (Ref. 13, by permission).

ference filter is used in the emission side. The main advantage of this system is that it can be easily and inexpensively constructed using off-the-shelf components.

4.5. Conclusion

The use of NIR-absorbing chromophores as labels offers many advantages in HPLC separations. Lower detection limits and lower background are major advantages. This research is still in its infancy, mostly due to the lack of availability of commercial NIR-absorbing labels. Significant advancement has been recorded in the development of appropriate semiconductor light sources, making this technique affordable and attractive. Covalent labeling with functionalized NIR fluorophores offers the greatest utility. When coupled with spectroscopic techniques that use semiconductor laser diodes as a light source, selective covalent labeling can have wide applications in all areas of HPLC analyses. It is expected that significant results will be reported in the near future.

4.6. Acknowledgment

This work was supported in part by a grant from the National Science Foundation (CHE-890456) and in part by a grant from the National Institutes of Health (Grant 1 RO 1 AI 28903-01A2).

4.7. References

1. L. W. Hershberger, J. B. Callis, and G. D. Christian, *Anal. Chem.* **51,** (1979) 1444.

2. D. B. Skoropinski, J. B. Callis, J. D. Sheldon Danielson, and G. D. Christian, *Anal. Chem.* **58** (1986) 2831.

3. K. Tanabe, M. Glick, B. Smith, E. Voigtman, and J. D. Winefordner, *Anal. Chem.* **59** (1987) 1125.

4. G. G. Guilbault, Ed., *Practical Fluorescence: Theory, Methods, and Techniques,* Marcel Dekker: New York (1973).

5. J. R. Lakowicz, *Principles of Fluorescence Spectroscopy,* Plenum Press: New York (1983).

6. H. G. Barth, W. E. Barber, C. H. Lochmuller, R. E. Majors, and F. E. Regnier, *Anal. Chem.* **58** (1986) 211R.

7. R. W. Frei, H. Jansen, and U. A. Th. Brinkman, *Anal. Chem.* **57** (1985) 1529A.

8. I. S. Krull, Ed., *Reaction Detection in Liquid Chromatography,* Dekker: New York (1986).

9. R. J. Hurtubise, *Phosphorimetry: Theory, Instrumentation, and Applications,* VCH: New York (1990).

10. A. Bhattacharyya, *Ind. J. Biochem. Biophys.* **23** (1986) 171.

11. F. V. Bright, *Anal. Chem.* **60** (1988) 1031A.

12. D. J. Fry, in *Rodd's Chemistry of Carbon Compounds,* 2nd Ed., S. Coffey, Ed., Elsevier: New York (1977).

13. K. Sauda, T. Imasaka, and N. Ishibashi, *Anal. Chem.* **58** (1986) 2649.

14. R. C. Benson, and H. A. Kues, *J. Chem. Eng. Data* **22** (1977) 379.

15. L. J. E. Hofer, R. J. Grabenstetter, and E. O. Wiig, *J. Chem. Phys.* **72** (1950) 203.

16. M. Matsuoka, Ed., *Infrared Absorbing Dyes,* Plenum: New York (1990).

17. C. C. Leznoff and A. B. P. Lever, Eds., *Phthalocyanines, Properties and Applications,* VCH: New York (1989).

18. K. Kamisaka, I. Listowski, J. J. Betheil, and I. M. Arias, *Biochim. Biophys. Acta* **365** (1974) 169.

19. E. Unger and G. Patonay, *Anal. Chem.* **61** (1989) 1425.

20. J. Hicks, and G. Patonay, *Anal. Instr.* **18** (1989) 213.

21. M. A. Roberson, D. Andrews-Wilberforce, D. C. Norris, and G. Patonay, *Anal. Lett.* **23** (1990) 719.

22. R. J. Williams, M. Lipowska, and G. Patonay, *Proc. XVI NOBCChE Meeting,* (in press).

5

Electrochemical Detection for Liquid Chromatography

Joseph Wang

Department of Chemistry
New Mexico State University
Las Cruces, New Mexico 88003

5.1. Introduction

An electrochemical detector uses the electrochemical properties of compounds for their determination in a flowing stream. Since the pioneering work of Kissinger[1] in the early 1970s, liquid chromatography with electrochemistry (LCEC) has become a well-established and versatile analytical technique. Nowadays, LCEC is commonly used in many clinical, environmental, or industrial laboratories. Highly sensitive and selective measurements of a wide variety of compounds (in various complex matrices) have thus been reported. In addition to their inherent sensitivity and selectivity, electrochemical detectors are characterized by a wide linear range (over 4–5 orders of magnitude), low dead volumes, fast response, and relatively low cost. While parameters such as current, potential, conductivity, or capacitance can be monitored by various electrochemical detectors, this chapter will focus primarily on the most popular constant-potential measurements. Several books provide the desired background information on electrochemical processes and instrumentation.[2–5]

In the following sections the principles, requirements, and applications of LCEC will be described.

5.1.1. Principles

Electrochemical detection works on many different principles. Constant-potential (amperometric or coulometric) measurements are the most widely used schemes. Such detection is used primarily to measure compounds that are electroactive. For this purpose, the potential of the working electrode is controlled to cause the

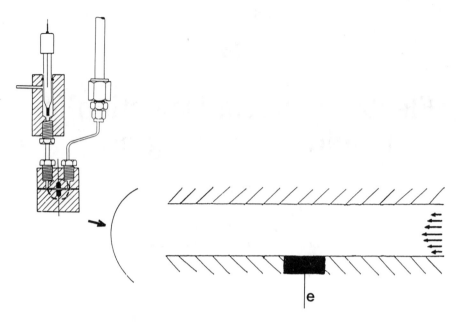

Figure 5.1.   Amperometric detection in a thin-layer channel.

chemical species (in the band flowing over the surface) to be reduced or oxidized (Figure 5.1). The applied potential represents the electrical energy that serves as the driving force of the redox process. Oxidation and reduction reactions require positive or negative potentials, respectively. The exact potential value is dependent upon the redox behavior of the compounds of interest (reflecting their tendency to gain or lose electrons). The current resulting from this electron-transfer process is related to the flux of material to the surface. Hence, the peak currents generated represent the concentration profiles of eluting compounds as they pass through the detector. Accordingly, the magnitude of the peak current serves as a measure of the concentration. The current peaks are superimposed on a constant background current (caused by redox reactions of the mobile phase). Larger background currents, expected at high potentials, result in increased (flow-rate-dependent) noise level. Increased noise level, associated with the presence of dissolved oxygen in the mobile phase, hampers the cathodic detection of reducible species. The signal-to-noise-ratio provides a measure of the minimum detectable quantity of the analyte. Readily oxidizable compounds can be detected at the range of picograms/injection. Since each class of compounds exhibits different electrochemical behavior (Figure 5.2), the operating potential can be exploited for selective detection (see the following discussion). In general, a lower potential is more selective, and a higher potential more universal. Thus, compounds undergoing redox reactions at lower potentials can be determined with greater selectivity.

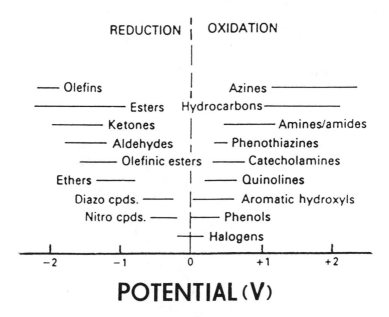

Figure 5.2. Electrochemical behavior of different organic functional groups.

Proper use of LCEC requires knowledge of the redox reactions and their dependence on the composition of the mobile phase. In most cases, a short cyclic voltammetry experiment can provide the desired information. Cyclic voltammograms of the compounds of interest can thus be rapidly obtained in several possible mobile phases. This type of experiment is performed in a batch cell containing a quiescent solution that resembles the one found in the LCEC detector. A more accurate (but time-consuming) approach would be the construction of hydrodynamic voltammograms (HDVs). This approach commonly serves for selecting the applied potential for the amperometric detection. The choice of the proper potential is dictated by the working and reference electrodes and the mobile phase composition and pH. Hydrodynamic voltammograms can be obtained by making repeated injections of the analyte solution while recording the current at different potentials. (The easiest way to accomplish this is by the flow injection method.) The resulting voltammograms have a wave (sigmoidal) shape, characterized by a half-wave potential and a limiting current region. Figure 5.3 shows HDVs for several biogenic amines. Such HDVs indicate the optimum operating potential. Although operation on the limiting current plateau region offers the highest sensitivity, a lowering of the potential (to the rising portion of the wave) can be used to improve the selectivity and lower the detection limit. For example, if compounds a, b, and g coelute, lowering of the operating potential to +0.45 V will allow selective monitoring of g. Comparison of HDVs for the sample peak and standard can also provide important information regarding the peak identity.

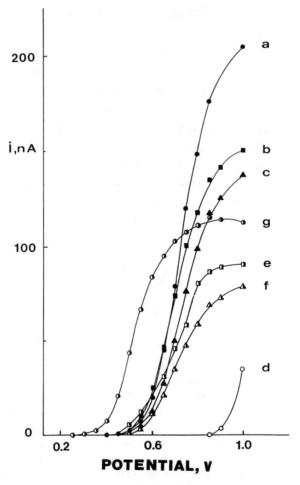

Figure 5.3. Hydrodynamic voltammograms of biogenic amines at a glassy carbon thin-layer detector: a, norepinephrine; b, L-dopa; c, epinephrine; d, tyrosine; e, dopamine; f, methyldopa; and g, homogentisic acid.

5.1.2. Theory

The use of theory to guide optimization efforts has helped with the design of powerful detectors. The measured current of an electrochemical detector is governed by the convective diffusion toward the electrode surface and ruled by the hydrodynamic conditions of the flow. For mass-transport-controlled (reversible) reactions, the current i is related to the flux of material to the surface, as described by the following equation:

$$i = nFAD \left(\frac{\delta C}{\delta x} \right)_{x=0,i}, \tag{5.1}$$

where n is the number of electrons transferred per molecule, F the value of the faraday, A the electrode surface area, D the diffusion coefficient, and $(\delta C/\delta x)_{x=0,t}$ the slope of the concentration–distance profile at the electrode surface. The electrode is usually operated in the limiting current region of the eluted species. Under this condition, the limiting current is given by:

$$i_D = \frac{nFADC}{\delta}, \tag{5.2}$$

where δ is the thickness of the diffusion layer. The latter is empirically related to the solution flow rate (U) via:

$$\delta = \frac{B}{U^\alpha}, \tag{5.3}$$

where B and α are constants for a given set of conditions, with α ranging between 0.33 to 1.0. By substituting Eq. (5.3) into Eq. (5.2), one obtains a general equation, commonly used to describe the limiting steady-state response of flow-through electrodes:

$$i_1 = nFAK_mCU^\alpha, \tag{5.4}$$

where K_m is the mass transport coefficient (D/B).

Another general equation for the limiting current of flow detectors is[6]:

$$i_1 = knFCD(Sc)^\beta \, b \, (R_e)^\alpha, \tag{5.5}$$

where k is a dimensionless constant, Sc the dimensionless Schmidt number $(v/D, v$ being the kinematic viscosity), b a measure of the electrode width, and R_e the dimensionless Reynolds number $(Ul/v, l$ being the characteristic length of the electrode). The values of α and β are determined by the hydrodynamic conditions, while k and b depend on the cell geometry. The values of these parameters are summarized in Table 5.1.

Although large electrodes give higher signals, the signal-to-noise ratio (S/N) is increased (and hence detection limits decrease) as the perimeter-to-area ratio increases.[7] Additional lowering of the detection limit can be achieved by breaking the electrode into pieces separated by an insulator.[8] The electrochemical noise (N) is the sum of various noise contributions. Depending on the frequency, several of these are proportional to the electrode area. Sources of the noise in electrochemical detection,

Table 5.1.   PARAMETERS OF EQ. (5.5)

Detector type	k	b	β	α
Thin layer	0.8	w	$1/3$	$1/2$
Tube	8.0	$l^{1/3}r^{2/3}$	$1/3$	$1/3$
Disk	3.3	r	$1/3$	$1/2$

and factors affecting the detection limits of LCEC, have been thoroughly examined and discussed by Weber and co-workers.[7]

5.2. Mobile Phase Requirements

The polarity, pH, ionic strength, and electroactivity of the mobile phase are all important for the success of amperometric detection. Such detection has been used mainly in reverse-phase and ion-exchange separations, where polar mobile phases are used, with an inert electrolyte added to make it conductive. In contrast, the mobile phase in normal-phase separations is often not polar enough to support the ionic strength required for the potential control. However, such difficulty can be addressed by using a large-volume wall-jet detector,[9] ultramicro working electrodes,[10] or via a postcolumn addition of an electrolyte solution.[11] It is often important to buffer the pH of the mobile phase because most redox reactions of biological compounds either release or consume protons. Electrolyte or buffer concentrations ranging from 0.01 to 0.1M are sufficient. As the mobile-phase composition or pH is important for the chromatographic separation and electrochemical detection, a compromise may be required in certain situations. Separations under more acidic conditions require higher oxidation potentials. The purity of the solvents and electrolyte used is critical to minimize the background current and the corresponding noise level. Similarly, the large background response associated with the presence of dissolved oxygen (when cathodic detection is performed at high potentials) can be minimized by deaerating the mobile phase, the use of on-line electrochemical or chemical oxygen removal, or with dual (series) electrode operation.

Depending upon the sample matrix, a cleanup procedure may be required to minimize electrode passivation problems. In particular, proteins or lipids are removed from biological samples to avoid their adsorption onto the detector surface.

5.3. Electrodes

The working electrode is the one at which the reaction of interest occurs. The choice of the working electrode is critical to the success of LCEC. Much of the developmental work on LCEC has thus been focused on new electrode systems. The working electrode material chosen for a given application should provide high sensitivity, selectivity, and stability. The following factors are thus considered: the kinetics of the electron-transfer reaction, the potential limits and background processes of the electrode in the mobile phase, and its physical and chemical resistance to this medium. While solid electrodes are commonly employed for oxidative work, mercury surfaces are advantageous for monitoring reducible species.

Solid electrodes include various types of carbon and different noble metals. Glassy carbon is the most widely used carbon electrode material. It is inert to common organic solvents, is impermeable to gas, and has a wide potential range (from -1.0 to $+1.3$ V). Before operation, glassy carbon should be carefully polished to a mirrorlike appearance using standard metallographic procedures. Electrochemical pretreatment and other reactivation schemes have proven useful in further enhancing the analytical capability of glassy-carbon-based detectors. Carbon paste electrodes, which use graphite powder mixed with organic binder (e.g., mineral oil), offer the advantages of low background current and noise levels, renewable surface, and low cost. A drawback of carbon paste electrodes is the tendency of the organic binder to dissolve in mobile phases containing an appreciable fraction (20–30%) of organic solvent. Among the various forms of porous carbon flow-through electrodes, reticulated vitreous carbon (RVC) appears to be especially suitable for LCEC applications.[12] Due to the slow equilibration of carbon surfaces at new potentials, the operation of carbon-based detectors requires 20–40 min of daily startup to achieve stable baselines.

Applications of gold and platinum electrodes have found increasing use for pulsed-amperometric detection of compounds (e.g., carbohydrates or alcohols) undergoing surface-catalyzed oxidations.[13] Such compounds may also be detected using nickel or copper working electrodes.[14,15]

Mercury as an electrode material offers the widest cathodic potential range, and it is thus commonly used for the detection of reducible species. Its anodic range, however, is very limited. Particularly attractive are detectors utilizing static mercury drops. The effluent is commonly introduced through a vertical or horizontal inlet jet into the space with the drop. Rapidly dispensed mercury drop electrodes have also been popular, particularly because of their renewable nature. The construction and operation of detectors based on mercury drop electrodes have been reviewed.[16] Mercury-coated gold or glassy carbon electrodes can also be employed.

The development of new electrode materials is generating considerable interest for LCEC. In particular, miniaturized and tailored surfaces can greatly enhance the power and scope of amperometric detection.

Microelectrodes, with at least one dimension in the micrometer domain, have attracted considerable attention in recent years.[17] When employed as flow detectors, such electrodes offer current density enhancement, reduced flow rate fluctuations, minimization of resistance effects, and a fast potential scanning capability. It is possible to couple many of the above advantages with larger currents through the use of arrays of multiple microelectrodes or of composite materials (consisted of active "islands" surrounded by an insulating region).[8] Kelgraf,[18] epoxy-impregnated RVC,[19] ultra-thin ring,[20] or microband[21] electrodes are good examples of composite or miniaturized detector surfaces for LCEC.

The use of chemically modified electrodes (CMEs) can also greatly benefit electrochemical detection in flowing streams.[22–24] Improved detection has been accomplished mainly through the use of electrocatalytic surfaces that accelerate electron-transfer reactions of many important analytes, or via coverage of the surface with a permselective layer to improve stability and selectivity. For example, by incorporating the catalyst cobalt phthalocyanine into carbon paste detectors, it is possible to lower (by several hundred millivolts) the potential required for monitoring hydrazines, thiols, or keto acids.[25–27] Coverage of the detector with size-exclusion celluloic films greatly simplifies complex chromatograms and offers protection against surface-active materials.[28] Ion-exchange inorganic coatings can be used to expand the scope of LCEC towards nonelectroactive cations.[29] It is possible also to biologically tailor the surface and thus to exploit the selectivity of enzymatic recognition processes.[30]

In addition to the working electrode, the flow cell contains the current-carrying auxiliary electrode (made of an inert conducting material, e.g., platinum) and a reference electrode (that provides a stable potential against which the potential of the working electrode is compared). Silver/silver chloride is the most commonly used reference electrode. The three-electrode system is connected to a potentiostat, which controls the potential of the working electrode while monitoring the resulting current.

5.4. Cell Design

Certain requirements must be fulfilled in designing electrochemical detectors for LCEC; these include high sensitivity, low dead volume, defined hydrodynamics, and ease of construction and maintenance. In addition, the reference and counter electrodes should be located on the downstream side of the working electrode, so that reaction products at the counter electrode or leakage from the reference electrode do not interfere with the working electrode. The most popular cells (which are also available commercially) are those in which the solution flows through a thin-layer channel[31] or onto a wall-jet electrode.[32] In thin-layer cells, a thin layer of solution flows parallel to the electrode surface in a rectangular channel (Figure 5.4). In the wall-jet design, the stream flows from a nozzle perpendicularly onto the electrode surface (Figure 5.5). Working electrodes with a diameter of 2–4 mm are commonly used in these configurations. The detectors are usually constructed from Kel-F or Teflon.

In addition to detectors with the effluent flowing by or onto the sensing element, it is possible to employ flow-through electrodes, with the electrode simply an open tube or porous matrix. The latter can offer a coulometric (high-conversion) detection, with increased signal and noise levels.

5.5. Detection Modes

The simplest and by far the most common detection scheme is the measurement of the current at a constant potential. Such constant-potential measurements have the advantage of being free of a double-layer charging the surface transient effects. As a result, extremely low detection limits—on the order of 1–100 pg (about 10^{-14} moles of analyte)—can be achieved. In various situations, however, it may be desirable to change (pulse, scan, etc.) the potential during detection.

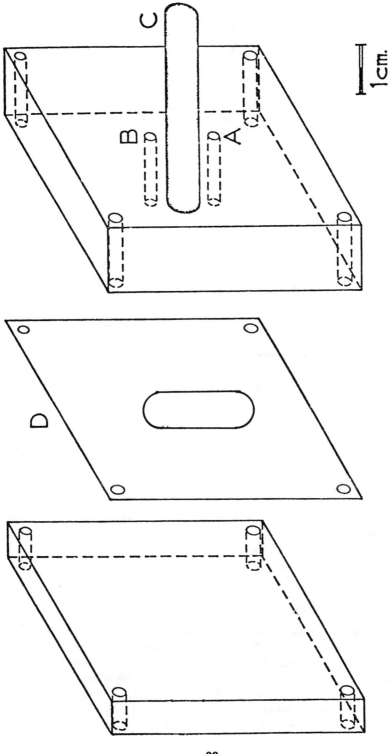

Figure 5.4. The thin-layer detector: A, inlet; B, outlet; C, working electrode; and D, Teflon spacer (the reference and auxiliary electrodes are placed in a downstream compartment; not shown).

99

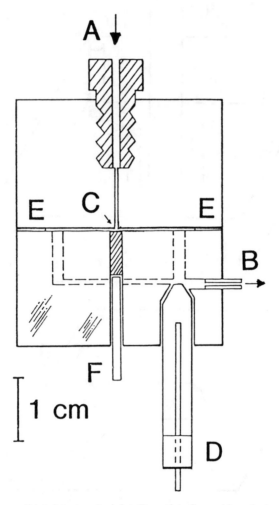

Figure 5.5. The wall-jet detector: A, inlet; B, outlet; C, working electrode; D, reference electrode; and E, Teflon spacer.

5.5.1. Pulsed Amperometric Detection

The development of pulsed amperometric detection (PAD), because of the efforts of Johnson and co-workers,[13,33,34] has greatly extended the scope and power of LCEC. This detection mode overcomes the problem of lost activity of novel metal electrodes associated with the constant-potential detection of numerous aliphatic compounds. For this purpose, PAD couples the process of anodic detection with cleaning and reactivation of the electrode. This is accomplished with a three-step potential waveform, combining anodic and cathodic polarizations (e.g., Figure

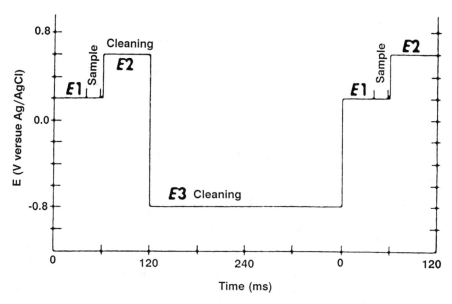

Figure 5.6.   Triple-pulse amperometric waveform.

5.6). Typically, such waveforms are executed at a frequency of 1–2 Hz. LC with PAD has thus been shown to be a highly selective, sensitive, and simple method for the determination of carbohydrates, alcohols, polyalcohols, amino acids, amines, aminoglycosides, and sulfur compounds. Gold and platinum working electrodes have been particularly useful for these LC/PAD applications. For a more detailed description of PAD, the reader is referred to recent reviews.[13,34]

5.5.2. Potential Scanning

Potential-scanning (voltammetric) detectors can increase the information content over that of fixed-potential (amperometric) operation. By rapidly recording numerous voltammograms during the elution, one obtains a three-dimensional detector response of the current against potential and time. Such addition of the redox potential selectivity can offer immediate identification of eluting peaks and helps in resolving chromatographically co-eluting components. Different approaches to swept-potential detectors based on square-wave voltammetry,[35–37] phase-sensitive voltammetry,[38] normal-pulse voltammetry,[39] adsorptive stripping voltammetry,[40] or coulostatic measurements[41,42] have been reported. The greater selectivity of potential-scanning detection is accompanied by higher detection limits (versus fixed-potential amperometry), owing to the additional background current associated with the potential scan. Advances in microelectrodes, dual-electrode detection, and data manipulation (background subtraction, derivatization) promise to address the higher detection limits.[43,44]

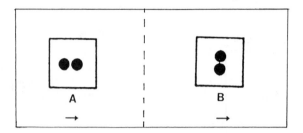

Figure 5.7. (a) Series and (b) parallel dual-electrode configurations.

5.5.3. Multiple-Electrode Detection

Multiple-electrode detection can also be employed to enhance the information content of LCEC. Different strategies, based primarily on dual-electrode detection, have been employed. For example, a more selective and sensitive detection can be achieved using the series configuration, where an upstream working electrode generates products that can be detected at the downstream electrode, held at a more favorable potential [Figure 5.7(a)]. Such configuration can also obviate the necessity of oxygen removal, through the downstream anodic detection of reducible species. Additional improvements in the power of the dual-series detection approach can be achieved utilizing an interdigitated microarray electrode, which offers higher collection efficiencies.[45] Sixteen simultaneous independent chromatograms have been obtained with a series array of 16 coulometric electrodes.[46] The electrodes of this series were held at incrementally higher potentials, so that compounds oxidized at the first electrode do not reach subsequent ones.

Dual electrodes in the (side-by-side) parallel configuration [Figure 5.7(b)] can offer other advantages. For example, significantly improved qualitative information can be achieved by holding these electrodes at different potentials and calculating the current ratios (of the simultaneously generated chromatograms). Such ratios provide real-time "fingerprints" of the eluting peaks. Additionally, highly useful information can be achieved by employing arrays of partially selective coated electrodes, in connection with equipotential operation and a chemometric approach.[47] A parallel-opposed dual-electrode configuration can also be employed to enhance the sensitivity through a redox-cycle amplification.[48] These and other developments in dual-electrode detection were reviewed by Roston et al.[49]

5.5.4. Differential Pulse Amperometry

Differential-pulse amperometry has been shown to improve the selectivity for compounds that react at potentials higher than other coeluting compounds.[50] For this purpose the potential of the working electrode is pulsed from a base potential, chosen around the half-wave potential of the analyte, and the difference in the current before and at the end of the pulse is measured (Figure 5.8). This operation is responsive only to compounds with redox potentials within the pulse width, hence

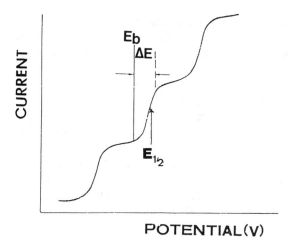

Figure 5.8. Differential-pulse detection. E_b is the base potential, ΔE is the pulse amplitude, and $E_{1/2}$ is the half-wave potential.

"filtering out" other potential interferences. Such a pulse technique has a larger background current and noise level than constant-potential measurements.

5.5.5. Tensammetric Detection

Nonelectroactive compounds that adsorb at a mercury electrode surface may be determined by tensammetry. In this mode, the analytical signal is obtained from changes (depression) in the double-layer capacitance associated with the adsorption process. Hence, this scheme is inherently sensitive towards surface-active compounds. For the same reason, calibration plots may exhibit nonlinearity for high concentrations. Theoretical and practical considerations of the design and operation of tensammetric detectors can be found in Ref. 51.

5.5.6 Improved Detection via Derivatization

Electrochemical, chemical, photolytic, and enzymatic derivatization schemes have been developed for improving the electrochemical detection of species with normally unfavorable redox properties.[52,53] Postcolumn electrochemical generation is a logical approach in connection with amperometric detection. Such generation of detectable species can be accomplished with negligible band broadening. For example, Kok et al.[54] reported the postcolumn generation of bromine in connection with the detection of difficult-to-oxidize phenolic ethers. The use of a series dual-electrode represents a logical extension of the on-line electrochemical derivatization approach. Postcolumn enzyme reactors hold great promise for selective derivatization of compounds.[55] Such an application has been documented for the determination of amino acids, bile acids, or urea. Several immobilized enzyme reactors can be placed in series, as was elegantly demonstrated for the determination of various saccharides in fermentation broths.[56] Detection of amino acids and amines has also been facilitated by chemical derivatization with *o*-phthaldehyde.[57] Similarly, non-electroactive peptides have been detected via the use of a postcolumn biuret reagent.[58] Chemical derivatization, with in situ generation of dithiocarbamate complexes, has been used for the detection of various metal ions.[59] The introduction of new tagging reagents will certainly expand the scope and power of LCEC.

Krull and co-workers[53,60] have designed novel postcolumn photolytic derivatizations for LCEC. The photolysis unit (mercury lamp) was placed directly, on line, between the column and the detector, generating electroactive species on irradiation. This concept was applied to the determination of various classes of analytes. Useful qualitative information has been obtained by switching the mercury lamp on and off and comparing the signals at two parallel working electrodes.

5.5.7. Conductivity and Potentiometric Detections

In addition to constant-potential detectors that are commonly used for trace organic analysis, conductivity and potentiometric detectors are often used to monitor ionic species in chromatographic effluents and other flowing streams.

Conductivity detectors are sensitive to ionic solutes in a medium of low conductivity. Because of the growing role of ion chromatography, such detectors are gaining increasing attention. Conductivity detectors offer simple design and operation, wide linear dynamic range (up to six orders of magnitude), universal response to ionic species, and small dead volumes (down to 0.5 μL). The cell consists of two electrodes, usually made of platinum, through which the column effluent flows. A constant alternating potential is applied to the electrode, and the resulting current is monitored. This current provides a measure of the solution conductivity. The temperature must be carefully controlled because the response is temperature dependent.

Using a mobile phase with high conductivity decreases the detector sensitivity. The technique of background conductivity suppression is normally used to cope with the problem of the effluent background. It involves the use of a second (suppressor) column, downstream from the analytical column. This removes the background electrolyte ions, leaving only the ions of interest as the major conducting species in the effluent. For example, if the effluent is a sodium hydroxide solution, a suppressor column that exchanges protons for the effluent sodium ions can be used. These protons react with the hydroxide ions to produce water. Such ion chromatography systems with conductivity detection are suitable for measuring alkali metal and alkaline earth cations, as well as chloride, phosphate, nitrite, and sulfate anions, in various matrices.

In addition to low-frequency conductivity measurements, high-frequency procedures can be used to monitor changes in the capacitance of the effluent.[61] Such procedures use flow cells with metallic electrodes that constitute the plates of a capacitor in a tuned circuit of a radio-frequency oscillator. When the solute passes between the plates, energy is adsorbed, resulting in a change of frequency. Two basic configurations of capacitance cells, planar and cylindrical, are employed. A linear response, over four orders of magnitude, is obtained (provided that the oscillators are stable). As with low-frequency conductivity detectors, good thermal stability is required. It is also possible to correct the background drift due to temperature changes using a microprocessor-based system. Detection limits around 150 ng can be achieved with such background correction.[62]

Finally, electrodes coated with a poly(3-methylthiophene) have been shown recently to monitor conductivity changes associated with the passage of inorganic ions in ion chromatographic analysis.[63]

Potentiometric detectors with ion-selective (or enzyme) electrodes, which find widespread use in automated flow analyzers, are less popular for monitoring chromatographic effluents. The major reasons for this are the high specificity, low sensitivity, and slow response of ion-selective electrodes. Two basic configurations of flow detectors with ion-selective electrodes are used: the flow-through electrode (such as tubular steady electrodes) and the cap design. The latter consists simply of an electrode probe fitted tightly with a cap, with an inlet and outlet for the flowing stream. The specificity problem can be addressed by titrating the eluting components with anions sensed by the detector, by using electrodes of lower selectivity or via multielectrode detection. For example, a copper-selective electrode was used for monitoring eluting amino acids.[64] The effluent was mixed with a copper ion solution, and changes in the level of the free copper ions, associated with the reaction with amino acids, were monitored.

Table 5.2. SELECTED BIOMEDICAL APPLICATIONS OF LIQUID CHROMATOGRAPHY WITH ELECTROCHEMICAL DETECTION

Analyte	Matrix	Working electrode	Detection electrode	Ref.
Acetaminophen	Urine	Glassy carbon	Dual electrode	65
Cardiac glycosides	Biofluids	Hanging mercury drop	Tensammetric	66
Catecholamines	Urine	Carbon paste	DC amperometry	67
Ceruloplasmin	Serum	Modified glassy carbon	DC amperometry	68
cis-Platinum	Plasma	Hanging mercury drop	Differential pulse	69
Chloramphenicol	Blood	Hanging mercury drop	DC amperometry	70
Diazepam	Serum	Hanging mercury drop	DC amperometry	71
Morphine	Blood	Glassy carbon	DC amperometry	72
Penicillamine	Plasma	Glassy carbon	DC amperometry	73
Phenothiazines	Serum	Glassy carbon	DC amperometry	74
Pterins	Biological samples	Glassy carbon	Dual electrode	75
Thiopurines	Plasma	Modified carbon paste	DC amperometry	76
Uric acid	Urine, serum	Mercury drop	DC amperometry	77

Table 5.3.   SELECTED ENVIRONMENTAL APPLICATIONS
             OF LIQUID CHROMATOGRAPHY WITH ELECTROCHEMICAL DETECTION

Analyte	Matrix	Working electrode	Detection scheme	Ref.
Al, Fe, Mn	Natural water	Glassy carbon	DC amperometry	78
Aromatic amines	River water	Carbon paste	DC amperometry	79
Benzidine	Effluent	Glassy carbon	DC amperometry	80
Carbamate pesticides	Water	Kelgraf	DC amperometry	81
Chlorinated phenols	Effluent	Carbon	DC amperometry	82
Methyl mercury	Fish	Mercury	DC amperometry	83
Organotin cations	Water	Gold/amalgam	Differential pulse	50
Phenols	Natural water	Kelgraf	DC amperometry	84
Polynuclear hydrocarbons	Water	Glassy carbon	DC amperometry	85
Selenium-containing compounds	Water	Gold/amalgam	Dual electrode	86

5.6. Applications

The popularity of electrochemical detection for liquid chromatography is evident
from the great number of applications reported over the past 20 years. Most applications of LCEC have been concerned with the measurement of readily oxidizable
molecules, such as biogenic amines, phenols, indoles, thiols, organic acids, or
aminophenols. Attention has also been given to measurements of reducible species
such as quinones, nitro, nitroso, diazo, or organometallic compounds. Pulsed amperometric detection at noble metal electrodes and the use of electrocatalytic modified electrodes has greatly extended the scope of LCEC towards aliphatic compounds. Many other classes of compounds have been determined via the use of
different derivatization schemes. There is no shortage of references to applications
of LCEC to fields such as environmental, clinical, industrial, food, or forensic
analyses. Because of its inherent sensitivity and selectivity, LCEC has been particularly popular in the biomedical and environmental fields. (The technique got its start
in neurochemistry and continues to be that of choice for the measurement of neuro-
transmitters such as dopamine.) Representative clinical and environmental applications of LCEC are listed in Tables 5.2 and 5.3, respectively.

5.7. Conclusions

The development of electrochemical detectors and detection strategies for liquid
chromatography continues to be a rapidly growing area of research. Further improvements in the scope, stability, and selectivity of such detectors may be required
to meet new challenges posed by complex samples and the use of new (capillary or
micropacked) columns. We anticipate that significant developments, particularly the

introduction of new tailored microelectrodes or of novel derivatizations schemes, will further enhance the power and utility of electrochemical detection.

5.8. References

1. P. T. Kissinger, L. Felice, R. Riggin, L. Pachla, and D. Wenke, *Clin. Chem.* **20** (1974) 992.

2. A. J. Bard and L. R. Faulkner, *Electrochemical Methods,* Wiley: New York (1980).

3. P. T. Kissinger and W. R. Heineman, Eds., *Laboratory Techniques in Electroanalytical Chemistry,* Marcel Dekker: New York (1984).

4. J. Wang, *Electroanalytical Techniques in Clinical Chemistry and Laboratory Medicine* VCH Publishers: New York (1989).

5. K. Stulik and V. Pacakova, *Electroanalytical Measurements in Flowing Liquids,* Ellis Horwood: Chichester (1987).

6. H. B. Hanekamp, P. Box, and R. W. Frei, *Trends. Anal. Chem.* **1** (1982) 135.

7. S. E. Weber and J. T. Long, *Anal. Chem.* **60** (1988) 903A.

8. D. E. Tallman and S. L. Petersen, *Electroanalysis* **2** (1990) 499.

9. H. Gunasingham, B. T. Tay, and K. P. Ang, *Anal. Chem.* **56** (1984) 978.

10. A. M. Bond, M. Fleischmann, and J. Robinson, *J. Electroanal. Chem.* **168** (1984) 299.

11. M. Lemar and M. Porthault, *J. Chromatogr.* **130** (1977) 372.

12. J. Wang and H. D. Dewald, *J. Chromatogr.* **285** (1984) 281.

13. D. C. Johnson and W. R. La Course, *Anal. Chem* **62** (1990) 589A.

14. W. Buchbenger, K. Winsaner, and C. Breitwiesser, *Z. Anal. Chem.* **315** (1983) 518.

15. W. Kok, U. Brinkman, and R. W. Frei, *J. Chromatogr.* **256** (1983) 17.

16. W. W. Kubiak, *Electroanalysis* **1** (1989) 379.

17. J. Wang, Ed., *Microelectrodes,* VCH Publishers: New York (1990).

18. J. L. Anderson, K. K. Whiten, J. D. Brewster, T. N. Ou, and W. K. Nonidez, *Anal. Chem.* **57** (1985) 1366.

19. J. Wang and B. Freiha, *J. Chromatogr.* **289** (1984) 79.

20. J. Bixler, M. Fifield, J. Poler, A. Bond, and W. Thormann, *Electroanalysis* **1** (1989) 23.

21. J. Zadeii, R. Mitchell, and T. Kuwana, *Electroanalysis* **2** (1990) 209.

22. J. Wang, *Anal. Chim. Acta.* **234** (1990) 41.

23. R. P. Baldwin and K. N. Thomsen, *Talanta* **38** (1991) 1.

24. E. Wang, H. Ji, and W. Hou, *Electroanalysis* **3** (1991) 1.

25. K. Korfhage, K. Ravichandran, and R. P. Baldwin, *Anal. Chem.* **56** (1984) 1514.

26. M. K. Halbert and R. P. Baldwin, *J. Chromatogr.* **345** (1985) 43.

27. L. M. Santoa and R. P. Baldwin, *Anal. Chem.* **54** (1986) 848.

28. J. Wang and L. D. Hutchins, *Anal. Chem.* **57** (1985) 848.

29. K. N. Thomsen and R. P. Baldwin, *Electroanalysis* **2** (1990) 263.

30. M. P. Connor, J. Wang, W. Kubiak, and M. R. Smyth, *Anal. Chim. Acta* **229** (1990) 139.

31. P. T. Kissinger, *Anal. Chem.* **49** (1977) 447A.

32. B. Fleet and C. J. Little, *J. Chromatogra. Sci.* **12** (1974) 747.

33. S. Hughes, P. L. Meschi, and D. C. Johnson, *Anal. Chim. Acta* **132** (1981) 1.

34. D. Autin-Harrison and D. C. Johnson, *Electroanalysis* **1** (1989) 189.

35. J. Wang, E. Ouziel, C. Yarnitzky, and M. Ariel, *Anal. Chim. Acta* **102** (1978) 99.

36. R. Samulesson, J. O'Dea, and J. Osteryoung, *Anal. Chem.* **52** (1980) 2215.

37. P. A. Reardon, G. E. O'Brien, and P. E. Sturrock, *Anal. Chim. Acta* **162** (1984) 175.

38. A. Trojanek and H. G. De Jong, *Anal. Chim. Acta* **141** (1982) 115.

39. W. Caudill, A. G. Ewing, S. Jones, and R. M. Wightman, *Anal. Chem.* **55** (1983) 1877.

40. L. Zhang and P. E. Sturrock, *Electroanalysis* **2** (1990) 289.

41. T. A. Last, *Anal. Chem.* **55** (1983) 1509.

42. R. K. Trubey and T. A. Nieman, *Anal. Chem.* **58** (1986) 2549.

43. C. E. Lunte, *LC-GC* **7** (1988) 492.

44. C. E. Lunte, T. H. Ridgeway, and W. R. Heineman, *Anal. Chem.* **59** (1987) 761.

45. A. Aoki, T. Matsue, and I. Uchida, *Anal. Chem.* **62** (1990) 2206.

46. W. R. Matson, P. G. Gamache, M. F. Beal, and E. D. Bird, *Life Sci.* **41** (1987) 905.

47. J. Wang, G. Rayson, Z. Lu, and H. Wu, *Anal. Chem.* **62** (1990) 1924.

48. S. McClintock, W. C. Purdy, and S. N. Young, *Anal. Chim. Acta* **166** (1984) 171.

49. D. A. Roston, R. E. Shoup, and P. T. Kissinger, *Anal. Chem.* **54** (1982) 1417A.

50. W. A. MacCrehan, *Anal. Chem.* **53** (1981) 74.

51. H. De Jong, W. Kok, and P. Bos, *Anal. Chim. Acta* **155** (1983) 37.

52. P. T. Kissinger, K. Bratin, G. C. Bratin, and L. A. Pachla, *J. Chromatogr. Sci.* **17** (1979) 137.

53. I. S. Krull, X. D. Ding, C. Selavka, and R. Nelson, *LC* **2** (1984) 214.

54. W. Th. Kok, U. A. Th. Brinkman, and R. W. Frei, *Anal. Chim. Acta* **162** (1984) 19.

55. L. Dalgard, *Trends Anal. Chem.* **7** (1986) 185.

56. G. Marko-Varga, *Anal. Chem.* **61** (1989) 831.

57. P. Leroy, A. Nicolas, and A. Moreau, *J. Chromatogr.* **282** (1983) 561.

58. A. Warner and S. G. Weber, *Anal. Chem.* **61** (1989) 2664.

59. A. M. Bond and G. C. Wallace, *Anal. Chem.* **55** (1983) 716.

60. L. Dou and I. S. Krull, *Anal. Chem.* **62** (1990) 2599.

61. S. Haderka, *J. Chromatogr.* **54** (1971) 357.

62. J. Alder, P. Drew, and P. Fielden, *Anal. Chem.* **55** (1983) 256.

63. Z. L. Xue, A. Karagzler, O. Ataman, A. Glal, A. Amer, R. Shabana, H. Zimmer, and H. Mark, *Electroanalysis* **2** (1990) 1.

64. C. Loscombe, G. Cox, and J. Dabziel, *J. Chromatogr.* **166** (1978) 403.

65. L. Hutchins-Kumar, J. Wang, and P. Tuzhi, *Anal. Chem.* **58** (1986) 1019.

66. H. De Jong, W. Voogt, P. Box, and R. Frei, *J. Liq. Chromatogr.* **6** (1983) 1745.

67. R. Riggin and P. Kissinger, *Anal. Chem.* **49** (1977) 2109.

68. J. Ye, R. O. Baldwin, and J. Schlager, *Electroanalysis* **1** (1989) 133.

69. F. Elfering, J. van der Vijgh, and H. Pinedo, *Anal. Chem.* **58** (1986) 2293.

70. J. Van der Lee, J. Van der Lee-Rijsbergen, U. Tjaden, and W. Van Bennekom, *Anal. Chim. Acta* **29** (1983) 149.

71. N. Thuaud, N. B. Sebille, M. H. Livertoux, and J. J. Besseine, *J. Chromatogr.* **282** (1983) 509.

72. M. W. White, *J. Chromatogr.* **178** (1979) 229.

73. I. C. Shaw, A. E. McLean, and C. Boult, *J. Chromatogr.* **275** (1983) 206.

74. J. E. Wallace, E. Shimek, S. Stavchansky, and S. Harris, *Anal. Chem.* **53** (1981) 960.

75. C. E. Lunte and P. T. Kissinger, *Anal. Chem.* **55** (1983) 1458.

76. M. Halbert and R. Baldwin, *Anal. Chim. Acta* **187** (1986) 89.

77. F. Palmisano, E. Desimoni, and P. Zambonin, *J. Chromatogr.* **306** (1984) 205.

78. Y. Nagoasa, H. Kawake, and A. M. Bond, *Anal. Chem.* **63** (1991) 28.

79. J. R. Rice and P. T. Kissinger, *Environ. Sci. Technol.* **16** (1982) 263.

80. R. M. Riggin and C. C. Howard, *Anal. Chem.* **51** (1979) 210.

81. R. M. Riggin and C. C. Howard, *Anal. Chem.* **52** (1980) 2152.

82. D. N. Armentrout, J. C. McLean, and M. W. Long, *Anal. Chem.* **5** (1979) 1039.

83. W. Holak, *Analyst* **107** (1982) 1457.

84. D. E. Weisshaar, D. E. Tallman, and J. L. Anderson, *Anal. Chem.* **53** (1981) 1809.

85. M. G. Khaledi and J. G. Dorsey, *Anal. Chim. Acta* **161** (1984) 201.

86. H. Killa and D. Rabenstein, *Anal. Chem.* **60** (1988) 2283.

6

Photothermal Detectors for High-Performance Liquid Chromatography

Chieu D. Tran

Department of Chemistry, Marquette University
Milwaukee, Wisconsin 53233

6.1. Introduction

Absorption measurement is probably the most widely used type of detection for high-performance liquid chromatography (HPLC). This is because the absorption technique can be used to detect virtually any compound, since all molecules absorb light, whereas other techniques such as fluorescence and refractive index suffer from such limitation as the lack of fluorophore in the sample or the similarity between the refractive index of the sample and the mobile phase. Unfortunately, in spite of its versatility, all absorption detectors currently available suffer from inherently low sensitivity. This is due to the fact that these detectors are based on transmission, which is known to be an insensitive measurement. It is therefore desirable to have a novel technique that can measure absorption with a much higher sensitivity than the currently available transmission measurements.

The availability of lasers makes it possible to observe and measure a variety of phenomena that would not be feasible with other light sources.[1-3] One such phenomenon is the photothermal effect.[4-22] In the case of the thermal lens [4-10] which is one type of photothermal effect, a sample is excited by a laser beam, which has a symmetrical intensity distribution (TEM_{00}). The nonradiative relaxation releases the absorbed energy in the form of heat. The heat generated is strongest at the center of the beam, because that is where the beam intensity is strongest. This creates a temperature gradient, which in turn produces a refractive index gradient and changes the radial intensity distribution of the laser beam. Thus, the optical energy absorbed can be sensitively determined by measuring the change in the intensity of the laser beam as it passes through the sample. The sensitivity of this thermal lens technique is higher than conventional transmission measurements because the signal intensity is directly proportional to the laser light intensity. The thermal lens technique, as well as photothermal techniques in general, may be used as versatile and sensitive absorption detectors for HPLC.

The goal of this chapter is not to review the photothermal techniques extensively but rather to demonstrate their advantages as compared to conventional HPLC absorption detectors, and briefly to review the current and future prospects of the photothermal HPLC detector. The chapter will begin with a description of the principles of the techniques, followed by specific applications and anticipated future developments.

6.2. Theory

The following theory is presented for the thermal lens technique. It can be easily modified to explain other related techniques such as photothermal deflection and photothermal refraction.[11,12,16-18]

A molecule is excited into either a vibrational or electronic excited state by the absorption of light. Subsequently the excited molecule releases the excitation energy in the form of heat via nonradiative relaxation processes. If the radiative processes of the molecule are negligible compared to the nonradiative processes, the heat generated equals the excitation energy.

The heat Q generated per unit length and unit time within the irradiated sample is expressed as[19-21]:

$$Q(r) = \frac{2(\ln 10)PC\epsilon e^{-2r^2/\omega^2}}{\pi\omega^2}, \tag{6.1}$$

where r is the distance from the beam center, ω the laser beam radius, ϵ the absorption coefficient of the absorbing species with concentration C, and P the laser power.

An increase in temperature ΔT can be calculated for any time during and after excitation, by the use of the appropriate Green's function for the heat equation that describes the situation. The time dependence of the temperature change within the sample can be calculated as:

$$\Delta T(r,t) = \frac{2(\ln 10)PC\epsilon}{\pi\omega^2\rho c_p} \int_0^\infty \frac{\exp\left(-\dfrac{2r^2/\omega^2}{1 + 2(t - \tau)/t_c}\right) d\tau}{1 + 2(t - \tau)/t_c} \tag{6.2}$$

where c_p is heat capacity of the sample, ρ the density, t the time after the laser beam onset, τ the excitation period, t_c the characteristic time constant, which can be calculated from the sample's density, heat capacity, thermal conductivity k, and beam radius,

$$t_c = \frac{\omega^2\rho c_p}{4k}. \tag{6.3}$$

The temperature changes during excitation can be obtained by integrating Eq. (6.2) from $\tau = 0$ to t, where $t \leq \tau$. After the excitation is turned off, $Q(r, t)$ equals zero, and the thermal lens strength starts to decrease. This is due to the dissipation of heat into the environment. In such a situation τ is taken as the upper limit for the integration in Eq. (6.2). The following discussion will be focused mainly on the buildup of the thermal lens and its maximum value.

The change in the refractive index induced by the temperature increase can be written as:

$$\Delta n = \Delta T \frac{dn}{dT},\tag{6.4}$$

where n is the refractive index.

The heat generated in Eq. (6.1) and the temperature change in Eq. (6.2) are generalized for any light source. For the case of the thermal lens measurement, an expression relating the effect of the refractive index change to laser beam intensity in the far field is needed. To obtain such an expression, approximations are required, since, at present, it is not possible to solve the integral in eq. (6.2) analytically.[19-21] Currently, there are two different approximations, namely the parabolic and the aberrant approximations.[19-21] In the parabolic model the thermal lens is assumed to have a parabolic refractive index distribution.[21] The aberrant model takes into account the aberrant nature of thermal lens.[19,20] This is accomplished by use of the diffraction theory of aberrations.[19,20] While the mathematical treatment for the parabolic model is relatively simple, it is based on relatively unrealistic conditions (i.e., no aberration in the thermal lens). Furthermore, experimental results have indicated that the aberrant model is more appropriate than the parabolic model. As a consequence, only the aberrant approximation is described here.[20]

In the aberrant model, the treatment of the thermal lens effect on the beam propagation is based not on the focal length of the thermal lens but rather on the phase shift at the input plane of the laser beam or the optical-path-length variation ϕ (r, t) induced by the thermal lens.[19,20] The induced phase shift is proportional to the sample length l and the difference in refractive index in the beam center $(r = 0)$ and at radius r. It can, therefore, be derived in terms of temperature increase as expressed in Eq. (6.4),

$$\phi(r,t) = \frac{dn}{dT} [T(0,t) - T(r,t)]l.\tag{6.5}$$

If the induced phase shift is small (i.e., $2\pi\phi/\lambda << 1$) and if approximations such as $Z_2 >> \omega$, $Z_2 >> $ are Z_1 are made (Z_1 and Z_2 are the distance from the sample to the beam waist and to the detector, respectively), the diffraction integral can be written as:

$$U_{bc}(t) = \frac{(2\pi)^{1/2} U_0 \omega \theta^{-2\pi Z_2 i/\lambda} i}{\lambda Z_2} \int_0^\infty \left(1 - \frac{2\pi\phi i}{\lambda}\right) e^{-(1+\xi t)u} \, du,\tag{6.6}$$

where U_{bc} is the complex phase and amplitude of the waves on the axis at the output plane at the position of the detector; U_o is the initial value of U_{bc}; and the ratio r^2/ω^2

is substituted by u. By using Eq. (6.2) for the temperature change, the diffraction integral will have the final form:

$$U_{bc}(t) = \tag{6.7}$$

$$K_0 \int_0^\infty \left\{ 1 - \frac{\theta i}{t_c} \int_0^t \left[1 - \exp\left(- \frac{2u}{1 + 2(t - \tau)/t_c} \right) \right] d\tau \right\} e^{-(1+\xi i)u} \, du.$$

All preintegral constants are included in K_0.

Equation (6.7) can be solved by performing the integration over u and then over τ. There is no need to assume the parabolic function for the temperature distribution, as in the case of parabolic model. The solution for the beam center intensity variation, $I(t)$, is found as the square of the absolute value of complex phase and amplitude of waves, $|U_{bc}|^2$. For $t \leq \tau$ it has the form:

$$I(t) = I(0) \left\{ 1 - \theta \arctan\left(\frac{2\xi}{3 + \xi^2 + (9 + \xi^2)t_c/2t} \right) \right. \tag{6.8}$$

$$+ \left[\frac{\theta}{2} \arctan\left(\frac{2\xi}{3 + \xi^2 + (9 + \xi^2)t_c/2t} \right) \right]^2$$

$$\left. + \left[\frac{\theta}{4} \ln\left(\frac{[(2 + t_c/t)(3 + \xi^2) + 6t_c/t]^2 + 16\xi^2}{(9 + \xi^2)^2(2 + t_c/t)^2} \right) \right]^2 \right\},$$

where $I(0)$ is the initial beam center intensity, $\xi = Z_1/Z_c$, and

$$\theta = - \frac{2.303 PA(dn/dT)}{\lambda k}. \tag{6.9}$$

In thermal lens experiments, θ is usually less than 0.1. Therefore, terms equal to θ^2 can be omitted. Under these conditions, the steady-state thermal lens, expressed in terms of the relative beam center intensity, is:

$$\frac{I(0) - I(\infty)}{I(\infty)} = -1 + \left[1 - \theta \arctan\left(\frac{2\xi}{3 + \xi^2} \right) \right]^{-1}. \tag{6.10}$$

Differentiate Eq. (6.9) with respect to ξ reveals that the thermal lens strength is maximum when the sample is positioned $Z_c(3)^{1/2}$ from the beam waist. Therefore, at $\xi = 3^{1/2}$ and for small θ values, the relative change in beam center intensity $\Delta I/I$ is linearly proportional to θ:

$$\frac{\Delta I}{I} = \theta \arctan(3^{-1/2}) = \frac{\theta}{1.91}. \tag{6.11}$$

Substitute Eq. (6.9) into Eq. (6.10) to obtain:

$$\frac{\Delta I}{I} = \frac{1.206 PA(dn/dT)}{\lambda k}. \tag{6.12}$$

The thermal lens technique has many advantages compared to conventional absorption methods. The most pronounced one is the capability for small volume samples and the ultrasensitivity. The former advantage stems from the unique characteristics of the laser, namely its high spatial resolution, which enables the laser beam to be focused to a very small spot (the size of the spot is limited by diffraction and thus can be as small as the wavelength of the beam). As a consequence, the thermal lens method is particularly suited for small volume samples such as the effluent of HPLC.

The thermal lens technique has ultrasensitivity because its signal intensity is, as depicted in Eq. (6.12), directly proportional to the excitation laser power. As described in the following section, its sensitivity is significantly higher than those of the conventional absorption methods. The relative change in beam intensity, when determined by conventional absorption techniques for a weakly absorbing species having absorbance A, is

$$\frac{\Delta I}{I} = 1 - 10^{-A} = 2.303A. \tag{6.13}$$

It is therefore clear from Eqs. (6.12) and (6.13) that the sensitivity of thermal lens techniques is relatively higher than that of conventional absorption methods. The sensitivity enhancement factor E is calculated from Eqs. (6.12) and (6.13) to be:

$$E = \frac{P(dn/dT)}{1.91\lambda k}. \tag{6.14}$$

According to this equation, when a 632.8 nm laser beam of 50 mW is used for the measurement, the sensitivity of thermal lens in CCl_4 is estimated to be 237 times higher than that of conventional absorption techniques.

6.3. Instrumentation

6.3.1. Thermal Lens Spectrometry

In general, thermal lens can be categorized into two types: single beam and double beam, or pump/probe.

In the single-beam system, a laser beam serves as the excitation (pump) as well as the monitor (probe). The systematic diagram of the system is shown in Figure 6.1. As illustrated, the sample is illuminated by a cw laser whose amplitude is modulated by either a mechanical chopper or an electronic shutter. A p-i-n photodiode, placed behind a pinhole, is used to measure the change in the laser beam center intensity as it passes through the sample and is being absorbed. A lens placed between the laser and the sample is used to facilitate the alignment and to optimize the thermal lens signal. It was found that a change in the distance between the lens and the sample or, strictly speaking, the distance between the laser beam waist and the sample, can increase, decrease, or even diminish the thermal lens signal. For example, no signal could be observed when the sample is placed at the beam waist. It has been theoretically predicted and subsequently verified by experimental results that the thermal lens signal intensity is maximum when the distance between the

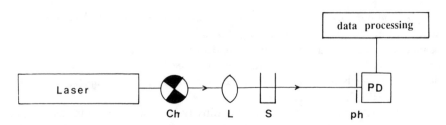

Figure 6.1. Schematic diagram of the single-beam thermal lens spectrometer: Ch, chopper; L, lens; S, sample; ph, pinhole; PD, *p-i-n* photodiode.

beam waist and the sample (i.e., Z_1) equals $3^{1/2}$ times the confocal distance Z_c.[18,19,21] From the definition $Z_c = \pi\omega_0^2/\lambda$, it is clear that Z_c can be calculated from the laser wavelength λ and the laser beam spot size at the beam waist ω_0. Since it is relatively time consuming to determine the ω_0 value accurately in practice, the alignment is performed by placing the lens on a translation stage and varying the Z_1 distance until maximum thermal lens signal is obtained. The thermal lens signal intensity was also found to be dependent on the distance between the sample and the detector (i.e., Z_2) and the aperture of the pinhole. However, for each Z_2 value, there is a certain size of the pinhole that gives a maximum and constant thermal lens signal for a given sample. Therefore, the selection of Z_2 and the aperture size is relatively simple; that is, it can be easily performed by initially deciding the Z_2 value (which normally is about 1.5 to 3.0 m) and subsequently adjusting the size of the pinhole until maximum signal is obtained.

Since a single laser beam is used for both excitation and monitoring, the data acquisition and analysis for the single-beam instrument is relatively time consuming. Generally, the beam center intensity of the laser beam is recorded as a function of time. Subsequent curve fitting of Eq. (6.8) will provide θ and t_c.[3] Alternatively, the θ value can also be calculated from the inverse of the intercept of the plots of $I_{bc}(t)/I_{bc}(0) - I_{bc}(t)$ versus $1/t$, where $I_{bc}(0)$ and $I_{bc}(t)$ are the beam center intensity at time $t = 0$ and t.[23] It is thus clear that, due to this time-consuming and cumbersome procedure, the single-beam system has not been extensively used as an HPLC detector.

The schematic diagram for a pump/probe thermal lens instrument is shown in Figure 6.2. Two laser beams are used in this apparatus: One that has relatively stronger power and is absorbed by the sample, is used as the pump beam, while the weaker one (normally not absorbed by the sample) is used as the probe beam. These two beams can be from two different lasers or from the same laser. Two lenses are used to focus the pump and the probe beam. The distance from the sample to the waist of the pump beam is different from that to the probe beam. It was found that the sample should be at the waist of the pump beam in order to receive maximum optical power (the power of the beam is inversely proportional to the square of the beam spot size). As with the case of the single-beam instrument, the distance from the sample to the probe beam waist should be equal to $3^{1/2}Z_c$.[22]

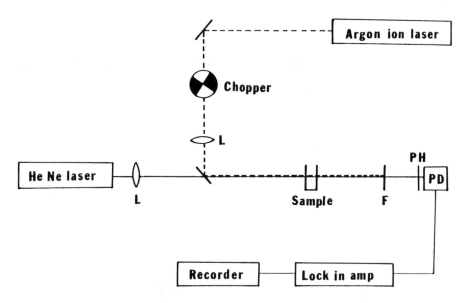

Figure 6.2.   Schematic diagram of the pump/probe thermal lens spectrometer. L, lens; F, pinhole; and PD, *p-i-n* photodiode. (Reproduced with permission from Ref. 6.)

Data acquisition for the double-beam instrument is simpler than with the single-beam apparatus. In the double beam, only the pump beam is modulated. As a consequence, a phase-locking device such as a lockin amplifier can be used for data acquisition.

6.3.2. Photothermal Deflection Spectrometry

Like the thermal lens technique, the photothermal deflection method (PTD) is based on the measurement of the change in the index of refraction of the optically heated region in the sample. The difference between the two techniques lies in the fact that in the thermal lens technique, the curvature of the thermally induced refractive index gradient is measured, whereas in the PTD, the deflection of the probe beam by this gradient refractive index is monitored. As a consequence, as shown in Figure 6.3, in the PTD the pump beam does not collinearly overlap with the probe beam but rather crosses the latter at a small angle (less than 1°) in the sample. The signal measured is, therefore, not the change in the beam center intensity of the probe beam as in the thermal lens method, but rather the deflection angle of the probe beam produced by the refractive index gradient. This deflection angle can be determined readily by use of a positioning sensor.

There is another type of the PTD technique that is based on the transverse configuration where the probe beam is propagating parallel to and very near (few nanometers) the sample surface; that is, the probing is accomplished in the thin layer

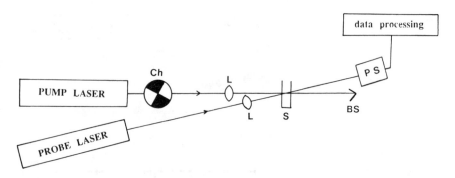

Figure 6.3. Schematic diagram of the photothermal deflection spectrometer: Ch, chopper; L, lens; S, sample; BS, beam stopper; PS, positioning sensor.

adjacent to the sample. This optical configuration is particularly suited for solid, opaque samples and also for mapping samples of poor optical quality.[9,11,12]

As expected, the sensitivity of the PTD technique depends on the overlapping length between the pump and probe beam within the sample. Since the crossing angle between the two beams is usually kept very small and the sample path length is very short (as in the case of the HPLC flow cell), the sensitivity of the PTD is expected to be similar to that of the thermal lens method. In fact, it has been determined recently that the sensitivities of the two techniques are comparable.[24] Additionally, the PTD has one advantage the thermal lens technique does not possess, namely, the pump laser used in the former technique need not have the intensity symmetrically distributed (i.e., TEM_{00}).

6.4. Applications

Application of the thermal lens technique to liquid chromatographic detection was performed as early as 1981 by Harris and co-workers with a single-beam apparatus. The inherent difficulty and time-consuming quality associated with the single-beam instrument necessitated the development of a novel data acquisition and analysis method. Specifically, they numerically fitted the thermal lens transient data during the 0.25 period while the sample was still cool. With this technique, they were able to obtain, using 190 mW laser power, a detection limit of 1.5×10^{-5} cm^{-1} for a 5 s response time.[25]

Relatively faster and easier data acquisition was accomplished in the following year by Morris and co-workers with a pump/probe instrument.[26] This is because in this instrument, the data acquisition and analysis can easily be done with a lockin amplifier. With the use of a 100 mW argon ion excitation laser beam, a detection limit of 2.0×10^{-6} absorbance units was achieved.[26]

Similar to other ultrasensitive techniques, effort in the thermal lens instrumentation has centered on the reduction of noise. A variety of noise reduction methods has been reported, including the detection of thermal lens signal not at the modulation frequency but rather at its second harmonic,[27] and the use of polarization to separate the excitation and the monitoring beam.[28,29] Of particular interest is the exploitation of the phase conjugation by means of a retroreflective array to provide not only the compensation for aberration but also double-pass advantage. The former advantage enables the reduction of baseline noise while the latter doubles the absorbance. Consequently, the signal-to-noise intensities for this instrument are improved by a factor of 4–6.[30]

Since flicker noise and the instability of lasers are the major contributors to the signal noise, it is possible to reduce the noise by performing the thermal lens measurement at high modulation frequency. In fact, Yeung and co-workers have demonstrated that, by use of an acousto-optic deflector to modulate a single laser beam at 150 kHz, the thermal lens signal produced can be measured with a lockin amplifier. With this apparatus, they were able to detect benzopurpurin in a microbore column at a level of 3 pg, which corresponds to 4.0×10^{-6} absorption units.[31]

As previously described, one of the advantages of the thermal lens technique is its ability to measure small volume samples. This advantage stems from the fact that the laser beam can be focused to a spot whose size is comparable to its wavelength. This advantage can be exploited to develop a novel type of thermal lens detector that is particularly suited for very small amounts of sample in a very small volume. To be able to achieve this objective, it was necessary, as shown in the instrument in Figure 6.4, to adopt an optical configuration in which the pump beam does not collinearly overlap but rather intersects the probe beam at a 90° angle (i.e., crossed beam thermal lens, CBTL). By use of this CBTL detector for microbore liquid chromatography, it was possible to detect amino acids at the femtomole level, which corresponds to only 50 molecules within the probe volume of 0.2 pL at the peak maximum.[32,33] The CBTL detector can also be constructed with a pulsed laser such as an excimer laser for excitation. The high peak power of the pulsed laser, however, produces some unwanted effects, such as the photodecomposition of the sample and the destruction of the flow cell. By using a relatively stable sample and a heavy walled flow cell, Sepaniak and co-workers have reported achieving a detection limit of 5.0×10^{-6} absorption units.[34]

In addition to the absorption measurement, the CBTL detector can also provide information on the refractive index (RI) of the chromatographic effluent. This is because the eluted sample produces two different effects on the probe beam; namely, it changes the unmodulated magnitude of the probe beam (due to the change in the refractive index of the effluent) and modulates the magnitude of the probe beam (by the thermal lens effect induced by absorption). As a consequence, by measuring the modulated and unmodulated component of the probe beam, one can simultaneously determine the refractive index and absorption of the eluted sample. In fact, by use of the instrument shown in Figure 6.5, Dovichi and co-workers were able to measure simultaneously the 4.0×10^{-6} change in the refractive index and the 6.3×10^{-6} change in the absorbance with their CBTL detector.[35]

Photothermal deflection (PTD), which is another type of photothermal technique, has also been used for liquid chromatographic detection. As previously

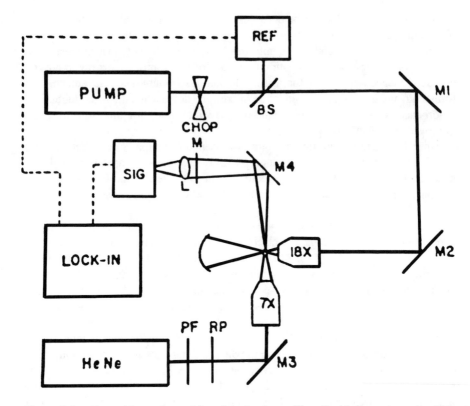

Figure 6.4. Crossed-beam thermal lens detector for capillary liquid chromatography: M1–M4, mirrors; PUMP, an argon ion laser; CHOP, mechanical chopper; BS, beamsplitter; REF, reference diode; 18× and 7×, 10- and 7-power microscope objective; HeNe, He-Ne laser; PF, polarizing filter; RP, quarter-wave retardation plate; M, mask formed by two razor blades; L, 50 mm focal length lens; SIG, signal photodetector. (Reproduced from Ref. 32 with permission.)

described, the sensitivity of the PTD, especially when a very short path length sample such as an HPLC flow cell is used, is comparable to that of the thermal lens technique. Furthermore, the instrumentation for the PTD technique is less restrictive than that for the thermal lens technique (i.e., the pump laser needs not have the TEM_{oo} mode). In spite of its potential, the application of the PTD technique to the analysis of solutions and liquid chromatographic detection is rather limited. The single study that used the PTD, shown in Figure 6.6 for microbore chromatographic detection, has demonstrated that it has comparable sensitivity to thermal lens techniques: A detection limit of 8.0×10^{-8} absorbance units was achieved with the use of a 1.2 W excitation laser power.[36]

A recent development, which is based on the thermal lens–circular dichroism measurement, is of particular importance because it enables, for the first time, the direct and real-time determination of the chirality and optical purity of chiral

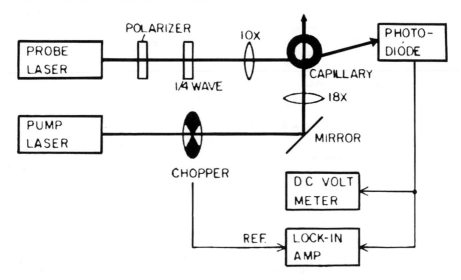

Figure 6.5. Schematic diagram of the detector for simultaneous measurement of refractive index and thermal lens. (Reproduced with permission from Ref. 34.)

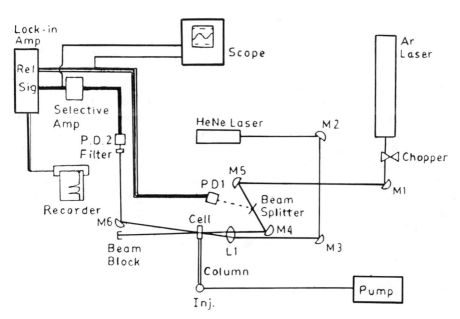

Figure 6.6. Schematic diagram for the photothermal deflection detector. M1–M6 are mirrors; L1 is a 100 mm focal length lens, PD1 and PD2 are photodiodes. (Reproduced from Ref. 36 with permission.)

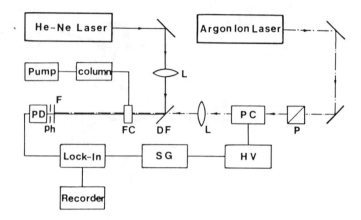

Figure 6.7. Thermal lens–circular dichroism chiral detector: P, prism polarizer; PC, Pockels cell; HV, high-voltage power supply; SG, signal generator; L, lens; DF, dichroic filter; FC, microflow cell; F, interference filter; PH, pinhole; PD, *p-i-n* photodiode. (Reproduced with permission from Ref. 38.)

effluent as it is eluted from a liquid chromatographic column.[37–39] This chiral detector is based on the measurement of circular dichroism of chiral effluents by the thermal lens effect. As shown in Figure 6.7, in the instrument the chromatographic effluent was sequentially excited by left circularly polarized laser light (LCPL) and right circularly polarized laser light (RCPL); the heat generated as a consequence of the sample absorption of the LCPL and RCPL was measured by the probe beam collinearly overlapping with the two excitation beams. A lockin amplifier was used to measure the thermal lens–circular dichroism (TL-CD) signal, which corresponds to the difference in the thermal lens signal produced by the LCPL and RCPL excitation beam. The sensitivity of this TL-CD apparatus is relatively higher than the conventional CD instrument, because the former is based on the measurement of the thermal lens effect, which is much more sensitive than the transmission measurement in the latter. A detection limit of 7.2 ng was achieved for (−)-tris(ethylenediamine)cobalt(III) (k' = 0.45) as well as for the (+)-tris(ethylenediamine)cobalt(III) (k' = 1.40) when these two enantiomers were chromatographically separated from the corresponding racemic mixture.[38] This limit of detection was found by using a 10 μL flow cell and having a 5 mm path length and a 6 mW excitation laser beam (λ = 514.5 nm) modulated at 2.0 Hz.[38]

6.5. Other Applications

The thermal lens and related photothermal techniques have also been used as a detector for flow injection analysis,[40,41] thin-layer chromatography,[42–46] gel electrophoresis,[47,48] and capillary electrophoresis.[49,50]

6.6. Future Prospects

It has been demonstrated that the thermal lens and related photothermal detectors have remarkable sensitivity. However, this ultrasensitivity does not necessarily mean that it cannot be further improved. In fact, in principle, the sensitivity of techniques can be further enhanced by instrumentation and/or chemical methods. Instrumentally, the detector sensitivity can be further improved by either increasing the signal intensity and/or decreasing the noise. It has been reported that the signal intensity of photothermal measurements can be substantially enhanced (up to 30 times) when the sample is placed inside the cavity of the probe laser.[13-15] Therefore, it is expected that up to at least a 30-fold improvement can be achieved when the HPLC flow cell is placed inside the cavity of the probe laser (i.e., intracavity photothermal HPLC detector).

One can also enhance the S/N by decreasing the noise. In such ultrasensitive techniques as the use of a photothermal detector, in addition to those contributed by electronic devices, noise is generated by solvent absorption. It is not possible to remove this background absorption using the photothermal detectors currently available because most of these instruments are based on the use of only a single wavelength for excitation. By use of the dual-wavelength (excitation) thermal lens spectrometer, it was demonstrated recently that background absorption by nickel glycinate at concentrations as high as $1.0 \times 10^{-2} M$ can be eliminated.[7,9,10] Thus, the S/N of the photothermal detector can be enhanced if it has the dual or multiwavelength capability to measure simultaneously the thermal lens signal at two or more wavelengths. If the multiwavelength photothermal detector is constructed using an acousto-optic tunable filter (AOTF), in addition to the aforementioned background discrimination, it will have extra advantages such as fast scanning, high resolution, reliability, and high reproducibility. The AOTF is an electronically driven optical dispersive device[51-54] based on the selective diffraction of light by acoustic waves. The AOTF in an all-solid-state device that has no moving parts, wide tuning range (from UV through visible to IR), high spectral resolution (few angstroms), and rapid scanning ability (μs).[51-54] The AOTF-based multiwavelength photothermal detector will, therefore, have the combined advantages of the photothermal technique and of the AOTF.

In addition to instrumentation improvement, the sensitivity of the photothermal detector can also be enhanced by chemical methods. As is apparent from Eq. (6.14), higher sensitivity can be achieved by performing the chromatographic separation in a solvent having high dn/dT and low thermal conductivity (k) values. Nonpolar solvents are good media for photothermal measurements owing to their high dn/dT and low k values. Water, which is the universal solvent for biochemical and biological samples, is the worst medium for photothermal techniques because it has very low dn/dT and high k values. In fact, we have calculated and experimentally verified that at the same laser intensity, thermal lens measurements in CCl_4 and n-pentane are 38 and 40 times higher than those in water.[8] It is, therefore, abundantly clear that higher photothermal signals can be obtained when the molar ratio of the nonpolar solvent in the mobile phase is higher than that of the polar solvent.

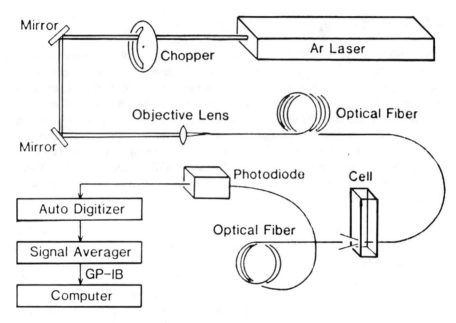

Figure 6.8. Fiber-optic-based thermal lens spectrometer. (Reproduced with permission from Ref. 55.)

It is particularly important to point out that, even without these improvements, photothermal detectors can still be fruitfully employed as an ultrasensitive and highly (molecular) specific detector for HPLC and related techniques. However, in spite of its potential and advantages, the present applications of photothermal techniques to chromatographic detection are still very limited. A variety of reasons might account for their limited use, but the most likely ones are probably the size of the instrument and the cost and reliability of the laser used. The former concern can be ameliorated because with the advances of electronics and fiber optics, it is possible to construct the thermal lens apparatus at a size comparable to the absorption detector. In fact, compact-size, fiber-optic-based thermal lens instruments such as the one shown in Figure 6.8 have been reported recently.[55–57] Concerns about the reliability and cost of the laser stem from information from the early years of the laser technology. The lasers that are currently available are relatively reliable and user friendly. Of particular interest is the rapid development in the field of semiconductor lasers. It is expected that diode lasers with UV, visible, and IR output will be readily available at very low price in the near future.[58] Taken together, it is evidently clear that the wide application of the photothermal techniques for HPLC detection is imminent.

6.7. Acknowledgment

The author is grateful to his dedicated and hard-working co-workers who are cited in the references. Acknowledgement is also made to Ms Patricia E. Kellogg for her competent graphical assistance. Financial support for this work is provided by the National Institutes of Health (grant number R01 RR06887-01).

6.8. References

1. G. M. Hieftje, J. C. Travis, and F. E. Lytle, *Lasers in Chemical Analysis,* Humana: Clifton, N.J. (1981).

2. D. S. Kliger, *Ultrasensitive Laser Spectroscopy,* Academic: New York (1983).

3. E. H. Piepmeier, *Analytical Applications of Lasers,* Wiley: New York (1986).

4. J. P. Gordon, R. C. C. Leite, R. S. Moore, S. P. S. Porto, and J. R. Whinnery, *J. Appl. Phys.* **36** (1965) 3.

5. J. M. Harris and N. J. Dovichi, *Anal. Chem.* **52** (1980) 695A.

6. C. D. Tran, *Anal. Chem.* **60** (1988) 182.

7. M. Franko and C. D. Tran, *Anal. Chem.* **60** (1988) 1925.

8. C. D. Tran and T. van Fleet, *Anal. Chem.* **60** (1988) 2478.

9. M. Franko and C. D. Tran, *Appl. Spectrosc.* **43** (1989) 661.

10. C. D. Tran and M. Franko, *J. Phys. E: Sci. Instrum.* **21** (1989) 586.

11. W. B. Jackson, N. M. Amer, A. C. Boccara, and D. Fournier, *Appl. Opt.* **20** (1981) 1333.

12. M. A. Olmstead, N. M. Amer, S. Kohn, D. Fournier, and A. C. Boccara, *Appl. Phys. A* **32** (1983) 141.

13. C. D. Tran, *Anal. Chem.* **58** (1986) 1714.

14. C. D. Tran, *Appl. Spectrosc.* **40** (1986) 1108.

15. C. D. Tran, *Appl. Spectrosc.* **41** (1987) 512.

16. C. D. Tran, *Analyst* **112** (1987) 1417.

17. N. Teramae and J. D. Winefordner, *Appl. Spectrosc.* **41** (1987) 164.

18. N. J. Dovichi, T. G. Nolan, and W. A. Wiemer, *Anal. Chem.* **56** (1984) 1700.

19. R. Vyas and R. Gupta, *Appl. Opt.* **27** (1988) 4701.

20. S. J. Sheldon, L. V. Knight, and J. M. Thorne, *Appl. Opt.* **21** (1982) 1663.

21. C. A. Carter and J. M. Harris, *Appl. Opt.* **23** (1984) 476.

22. T. Berthoud, N. Delorme, and P. Mauchien, *Anal. Chem.* **57** (1985) 1216.

23. M. Franko and C. D. Tran, *Anal. Chem.* **61** (1989) 1660.

24. Y. Yang and T. V. Ho, *Appl. Spectrosc.* **41** (1987) 583.

25. R. A. Leach and J. M. Harris, *J. Chromatogr.* **218** (1981) 15.

26. C. E. Buffett and M.D. Morris, *Anal. Chem.* **54** (1982) 1824.

27. T. K. J. Pang and M. D. Morris, *Anal. Chem.* **56** (1984) 1467.

28. T. K. J. Pang and M. D. Morris, *Appl. Spectrosc.* **39** (1985) 90.

29. Y. Yang, S. C. Hall, and M. S. De La Cruz, *Anal. Chem.* **58** (1986) 758.

30. T. K. J. Pang and M. D. Morris, *Anal. Chem.* **57** (1985) 2700.

31. K. J. Skogerboe and E. S. Yeung, *Anal. Chem.* **58** (1986) 1014.

32. T. G. Nolan, B. K. Hart, and N. J. Dovichi, *Anal. Chem.* **57** (1985) 2703.

33. T. G. Nolan and N. J. Dovichi, *Anal. Chem.* **59** (1987) 2803.

34. C. N. Kettler and M. J. Sepaniak, *Anal. Chem.* **59** (1987) 1736.

35. D. J. Bornhop and N. J. Dovichi, *Anal. Chem.* **59** (1987) 1632.

36. T. W. Collette, N. J. Parekh, J. H. Griffin, L. A. Carreira, and L. B. Rogers, *Appl. Spectrosc.* **40** (1986) 164.

37. C. D. Tran and M. Xu, *Rev. Sci. Instrum.* **60** (1989) 3207.

38. M. Xu and C. D. Tran, *Appl. Spectrosc.* **44** (1990) 962.

39. M. Xu and C. D. Tran, *Anal. Chem.* **62** (1991) 2467.

40. R. A. Leach and J. M. Harris, *Anal. Chim. Acta* **164** (1984) 91.

41. R. Yang and R. E. Hairrell, *Anal. Chem.* **56** (1984) 3002.

42. T. I. Chen and M. D. Morris, *Anal. Chem.* **56** (1984) 19.

43. T. I. Chen and M. D. Morris, *Anal. Chem.* **56** (1984) 1674.

44. T. Matsujima, A. Sharda, L. B. Lloyd, J. M. Harris, and E. M. Eyring, *Anal. Chem.* **56** (1984) 2977.

45. K. F. Fotiou and M. D. Morris, *Anal. Chem.* **40** (1986) 700.

46. C. D. Tran, *Appl. Spectrosc.* **41** (1987) 512.

47. K. Peck and M. D. Morris, *Anal. Chem.* **58** (1986) 506.

48. B. J. Jager, R. J. G. Carr, and C. R. Goward, *J. Chromatogr.* **472** (1989) 331.

49. C. W. Earle and N. J. Dovichi, *J. Liq. Chromatogr.* **12** (1989) 2575.

50. A. E. Bruno, A. Paulus, and D. J. Bornhop, *Appl. Spectrosc.* **45** (1991) 462.

51. T. Yano and A. Watanabe, *Appl. Opt.* **15** (1976) 2250.

52. I. C. Chang, *SPIE Acousto-Optics* **90** (1976) 12.

53. I. C. Chang, *Opt. Eng.* **20** (1981) 824.

54. A. Sivanayayam and D. Findley, *Appl. Opt.* **23** (1984) 4601.

55. K. Nakanishi, T. Imasaka, and N. Ishibashi, *Anal. Chem.* **59** (1987) 1550.

56. T. Imasaka, K. Nakanishi, and N. Ishibashi, *Anal. Chem.* **59** (1987) 1554.

57. F. Charbonnier and D. Fournier, *Rev. Sci. Instrum.* **57** (1986) 1126.

58. T. Imasaka and N. Ishibashi, *Anal. Chem.* **62** (1990) 363A.

HPLC Detection Using Fourier Transform Infrared Spectrometry

Victor F. Kalasinsky

Department of Environmental and Toxicologic Pathology
Armed Forces Institute of Pathology
Washington, DC 20306

Kathryn S. Kalasinsky

Division of Forensic Toxicology
Office of the Armed Forces Medical Examiner
Armed Forces Institute of Pathology
Washington, DC 20306

7.1. Introduction

7.1.1. Background

Infrared (IR) spectroscopy is used to study molecular vibrations, sometimes as a probe of internal energies and sometimes as a means of identifying a chemical species. The latter, of course, is important for various analytical applications, and the reader is referred to a wealth of books, chapters, and monographs detailing these applications.[1-6] The purpose of the current discussion is to describe and summarize the combination of high-performance liquid chromatography (HPLC) and infrared spectroscopy with a necessary emphasis on Fourier transform infrared (FT-IR) spectroscopy. Because all but the very simplest compounds have infrared absorptions, infrared spectroscopy has the potential to be a "universal" detector for HPLC. There are certain difficulties in configuring such a system, but results obtained thus far in a number of laboratories suggest a promising future for HPLC/FT-IR.

The desire to identify a chromatographically separated compound drives the coupling of chromatography and spectrometric techniques. The combination of gas chromatography and mass spectrometry (GC/MS) has become a commonly used and highly sensitive tool in the analytical arsenal.[7] GC/FT-IR provides complementary data for positively identifying an unknown,[8] but until recently, routine GC/FT-IR measurements were two to three orders of magnitude less sensitive than those of GC/MS.[9,10]

A rather obvious extension of GC/MS and GC/FT-IR was that of interfacing other chromatographic methods to the spectrometric techniques. Consequently, HPLC/MS systems based on a variety of different interfaces are commercially available.[11,12] Activities concerned with making highly sensitive, user-friendly HPLC/FT-IR systems available are also underway. Early HPLC/FT-IR interfaces utilized flow cells in a fashion analogous to more conventional UV–visible and fluorescence-based HPLC detectors. Problems with such a system for infrared detection include the absorption of the solvent and the various limitations of the optical materials needed to construct infrared cells; consequently, solvent-elimination techniques were applied to HPLC/FT-IR with greater ultimate success.

The difficulties and progress in efforts to develop suitable interfaces of both types will be described in this chapter. The following two introductory sections deal with considerations that are fundamental to infrared spectroscopy and emphasize its "mismatch" with HPLC. Subsequent sections address the specifics of flow-cell and solvent-elimination HPLC/FT-IR interfaces, other related chromatography-infrared techniques, and an evaluation of the "current" status of HPLC/FT-IR. This review is not meant to be exhaustive, because many of the details of early experimental work can be found in other articles.[13–20] Instead, we hope to provide a view of the realistic expectations for utilizing HPLC/FT-IR routinely in the analytical laboratory.

7.1.2. FT-IR Spectrometry

Dispersive infrared spectroscopy with instruments using prisms and eventually gratings as the dispersive elements has been used extensively for identifying chemical compounds, but many of today's very demanding applications are only feasible with Fourier transform infrared (FT-IR) techniques. FT-IR instruments became commercially available for spectroscopy in the late 1960s due in part to the emergence of minicomputers that could be dedicated to the task of performing the complicated mathematical Fourier transformation, and advances in computer-related technologies in the ensuing years have contributed significantly to the development of modern FT-IR spectrometry.

Most FT-IR instruments utilize the Michelson interferometer in which a beamsplitter divides the source output and directs one beam to a fixed mirror, while the other beam travels to a movable mirror. (See Figure 7.1.) The light reflected back from these two mirrors is recombined at the beamsplitter and sent to a detector. If the two optical pathways are of equal distance (and unobstructed), then there is constructive interference in the recombined beam. If the movable mirror is reposi-

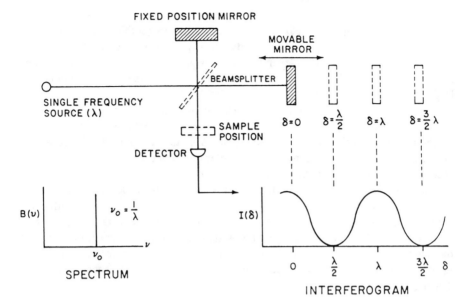

Figure 7.1. Optical diagram of a simplified Michelson interferometer. (Used with permission, Ref. 21.)

tioned to introduce an appropriate path-length difference, then it is possible to observe destructive interference. The extent of the constructive and destructive interference in the recombined beam depends on both the wavelength(s) of the light emanating from the source and the instantaneous optical path-length difference between the two beams in the interferometer.[5,21–23]

The interference pattern measured in an FT-IR instrument is known as an "interferogram" and consists of intensity variations as a function of the displacement of the movable mirror. A Fourier transformation converts the interference pattern of the different wavelengths of light into a more conventional spectrum, which displays intensity versus wavelength (or wave number). Since dispersive and FT-IR spectrometers ultimately give the same type of spectrum, it is worthwhile to mention briefly the advantages of FT-IR spectroscopy.[5,22]

The multiplex (or Fellgett's) advantage refers to the ability to record information simultaneously for the entire spectral range of interest. In a dispersive instrument, a grating (or prism) must be rotated so that narrow segments (spectral slit width) of the wavelength spectrum fall on the detector at any given time; information from these segments is recorded sequentially. In an FT-IR spectrometer, all the wavelengths are incident on the detector at all times; so spectra of comparable quality (comparable signal-to-noise ratio, S/N) can be collected in a fraction of the time. Therefore, if equal measurement times are used in dispersive and FT-IR experiments, the latter will give a much larger S/N.

The throughput (or Jacquinot's) advantage relates to the increased amount of light that arrives at the detector in an FT-IR instrument. The extent of this advantage depends heavily on specific instrument design parameters (both in dispersive and in FT-IR instruments), but generally the advantage goes to the latter. With infrared detectors, the noise level is independent of the signal level (unlike photomultiplier tubes in the UV–visible range); so an increased signal at the detector produces a larger signal-to-noise ratio.

The frequency accuracy in an FT-IR instrument is generally referred to as Connes's advantage. By using a helium–neon laser wavelength as a reference, the position of the movable mirror can be measured quite accurately. This accuracy in the distance (or mirror position) term in an interferogram translates into a frequency accuracy of 0.01 cm^{-1} in an infrared spectrum.

There are also other, more subtle advantages such as those dealing with reduced stray light and the high modulation frequencies, which will not be dealt with here, but they contribute to the fact that in an otherwise comparable situation, the signal-to-noise ratio (S/N) in an FT-IR spectrum is greater than that in a spectrum obtained by using a dispersive instrument. If measurement time is a constraint, the same factors allow FT-IR spectra to be collected in much less time than comparable dispersive spectra.

7.1.3. Design Considerations for HPLC/FT-IR

In designing and developing an interface between two dissimilar techniques, it is important to ensure that there is minimal degradation of the operating efficiencies of the techniques being considered. An understanding of the factors critical to each technique is necessary, and a certain amount of design flexibility is required if acceptable compromises are to be made. In terms of the compatibility of HPLC and FT-IR spectroscopy, two fundamental problems arise. Many common HPLC solvents exhibit strong infrared absorptions, and this is especially true for the polar solvents used for reversed-phase separations. Second, infrared window materials like alkali halide salts are incompatible with aqueous mobile phases; so it is important to choose optical components very carefully. Other materials are less sensitive to water but might suffer from a limited useful optical range, greater expense, or even inherent toxicity.

In an early article concerned with HPLC/FT-IR, Vidrine outlined factors to be considered in designing an interface.[24] Vidrine's list included items that apply to any desirable HPLC detector—it must be highly sensitive to analytes, compatible with all types of samples, compatible with all solvent systems, nondestructive to the analytes, able to provide real-time chromatograms, and compatible with flow and gradient programming. Two additional factors that are specific to spectroscopic detection were also given. These are the ability to monitor several spectral features (such as infrared bands) simultaneously and the ability to store complete spectra.

The experience of the intervening 12 years suggests that compatibility with analytical ($\geq$ 4-mm i.d.) and microbore ($\leq$ 1-mm i.d.) columns, the ability to conduct on-line or off-line experiments, the ability to adapt to changes in the conduct of the chromatographic experiment, and a compact size be added to the list as well. Of course, implicit in any discussion of an analytical procedure is the desire that the equipment be simple to operate and maintain.

In applying this extensive collection of criteria to infrared spectroscopy, some of them are satisfied simply by the nature of the infrared experiment, but others pose serious problems. For example, sensitivity and compatibility with samples are not problems while, on the other hand, aqueous (and other polar) solvents and buffers could contribute more bands (and more intense bands) to overlap the infrared spectrum of the analyte. While none of the interfaces described to date has success-fully incorporated all of the desirable features, significant progress has been made in designing functional HPLC/FT-IR systems. As with most analytical techniques, tradeoffs have to be made in HPLC/FT-IR, and the shortcomings of a given design have generally led to improvements in subsequent generations of HPLC/FT-IR interfaces. In the ensuing sections, we will describe the various approaches to combining HPLC and FT-IR spectroscopy and discuss the extent to which they satisfy design criteria and solve specific problems.

7.2. Flow-Cell Techniques for HPLC/FT-IR

7.2.1. Analytical-Column HPLC

In what appears to be the first literature reference to HPLC/FT-IR as a viable technique, a conventional infrared flow cell was connected to the effluent of a normal-phase, analytical-column HPLC system.[25] The sample load was quite high, and the analyte spectra were obtained only after spectral subtraction, but the concept was demonstrated. An automated flow-cell HPLC/FT-IR interface was described in 1978, and it utilized software originally developed for "lightpipe" (flow-cell) GC/FT-IR experiments,[26] which provided real-time infrared chromatograms. Fig-ure 7.2 shows a plot of infrared absorption versus time, and by monitoring five different frequency regions (3100–3000, 2990–2850, 1470–1440, 1100–1040, and 680–660 cm^{-1}), five infrared chromatograms were generated. It is clear that base-line resolution was not achieved in this GPC separation of a silicone oil, a paraffin oil, and benzene, but variations in peak shapes in the five infrared "windows" indicate where two of the components overlap in elution time. An alternative method for displaying the same information is a three-dimensional "stacked plot," such as that shown in Figure 7.3. The sequentially collected data files, which represent the time axis, are indicated by the numbers along the right side of the figure.

An additional feature of some importance in Figure 7.3 is the lack of spectral information in the region between 700 and 800 cm^{-1}. Over this region, the carbon tetrachloride solvent in the 0.2-mm-path-length, 1.5 μL flow cell completely ab-sorbs the light and makes it impossible to observe any solute peaks that might also be in the region. Even with this discontinuity, identifiable spectra of each compo-nent of the mixture were obtained. It is easy to imagine, though, that solvents with more complicated spectra can present problems in a flow cell, and proper choice of

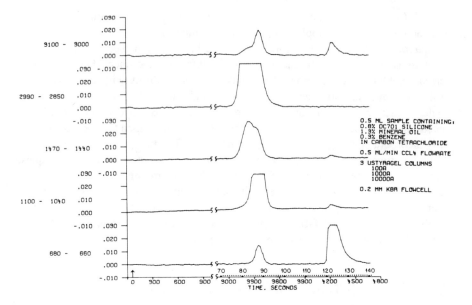

Figure 7.2. Infrared chromatograms for a GPC separation of a paraffin oil, a silicone oil, and benzene (elution order) in carbon tetrachloride. (Used with permission, Ref. 24.)

solvent is necessary to accommodate infrared detection in the GPC and normal-phase HPLC experiments described therein.

A more thorough discussion of the limitations and applications of this system was subsequently presented.[24,26] Solvent systems were considered in detail, and reference spectra of many different solvents for flow-cell HPLC/FT-IR were shown.[24] Most common solvents exhibit intense absorption bands over wide frequency regions, many of which overlap regions where elutes would be expected to absorb. As a general rule, it can be said that the more polar solvents tend to have stronger infrared absorptions than the less polar solvents. So while the question of compensating for solvent absorption is a significant consideration for normal-phase HPLC, it is a critical limitation for reverse-phase, HPLC especially with aqueous solvents.

Other applications of normal-phase HPLC/FT-IR using flow-cell detection have been reported.[28-30]. Johnson and Taylor used Freon 113 (1,1,2-tri-chlorotrifluoroethane) as the mobile phase in silica columns to separate jet fuel, coal-derived process solvents, and model compound mixtures for the process solvents to demonstrate the feasibility of semipreparative, analytical, and microbore HPLC/FT-IR.[31] Freon 113 was used because of its low polarity and because its midinfrared vibrational bands are all between 1300 and 700 cm^{-1}. This latter feature "opens" the important C–H stretching, C = O stretching, and aromatic-ring stretching regions even for flow-cell path lengths up to 1 mm. Taylor and co-workers extended the study of the process solvents by first using a semipreparative

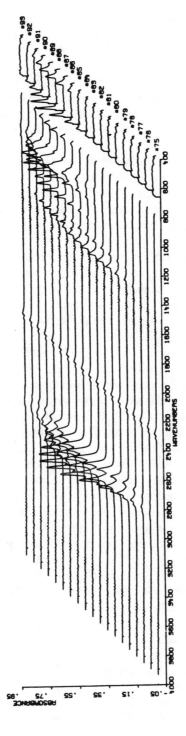

Figure 7.3. Infrared spectra from the separation in Figure 7.2 in order of acquisition during elution of the paraffin oil and silicone oil. (Used with permission, Ref. 24.)

133

method to partition materials into groups according to their polarities. Subsequent analytical HPLC was performed on the "intermediate polar"[28] and "polar"[29] fractions by using polar amino cyano (PAC) columns and chloroform or chloroform/acetonitrile (98:2) mixtures, respectively, as the mobile phases.

In the latter study, the addition of the polar modifier facilitated the separation of the analytes in what is still essentially a normal-phase experiment. In both cases, deuterated chloroform (CDCl$_3$) was used in the mobile phase for the HPLC/FT-IR work in order to minimize spectral overlap between solvent and analyte. The effect of deuteration is to shift bands involving carbon–hydrogen vibrations to lower frequencies (see Figure 7.4) by a factor approximately equal to the square root of the hydrogen–deuterium mass ratio; as in the case of Freon 113, this leaves the C–H stretching region open for "universal" detection of hydrocarbons. Results for analyses of nitrogen heterocycles and aromatics expected in coal-derived process solvents using CHCl$_3$ were acceptable,[30] but detection limits were better with the deuterated solvent.

Saunders and Taylor recently described the determination of the degree of nitration of cellulose nitrates by using gel permeation chromatography with flow-cell FT-IR detection.[32] They recognized that tetrahydrofuran (THF) has minimal absorption between 1600 and 1800 cm^{-1}, the same region where the O–N–O stretching vibrations of cellulose nitrate appear. The degree of nitration was determined by analyzing the bands in this region using the second derivative. A number of propellants was studied, and it was possible to use the technique to compare the degree of nitration as a function of molecular weight in the polymer.

The problems associated with solvent absorption can be minimized in normal-phase HPLC and GPC, as indicated in the examples given, but this task is much more complicated in reverse-phase HPLC, especially if aqueous mobile phases are used. With significant amounts of water, flow-cell path lengths must be extremely short (0.1–0.5 mm), and the flow cell must be impervious to water. One approach to flow-cell reverse-phase HPLC/FT-IR was demonstrated and refined by Taylor and co-workers.[33,34] In this system, the aqueous effluent was mixed with carbon tetrachloride, and the mixture was sent to a coil of tubing, which served as an extraction vessel. Once it passed through the extraction coil, the solvent mixture was directed into a membrane separator (Figure 7.5), which consisted of a triple-layer membrane sandwiched between two grooved stainless steel plates fitted with the appropriate inlet and outlet connectors. The membrane was constructed with an inner layer of 0.2-μm pore PTFE and outer layers of 1-μm pore PTFE supported on polypropylene, and its hydrophobic nature allowed only the nonpolar organic phase to pass through. Once the aqueous and organic (CCl$_4$) solvent streams were separated in this way, solutes were carried to a conventional flow-cell interface by the CCl$_4$, which, as we have already indicated, is a very useful solvent for infrared spectroscopy. The volumes of the flow cell and tubing in the system were minimized to try to maintain chromatographic resolution, and it was determined that the 32-μL volume of the membrane separator was responsible for the moderate band broadening observed in the study. With flow rates of 1 mL/min, it is unlikely that this volume could be reduced significantly while maintaining efficient phase separation.

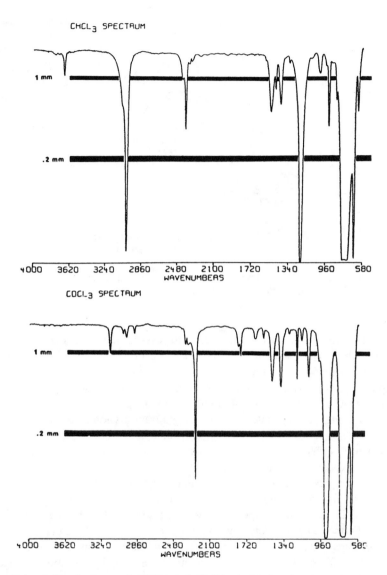

Figure 7.4. Infrared spectra of $CHCl_3$ and $CDCl_3$, indicating the absorption bands that interfere with solute spectra in 0.2- and 1-mm-path-length flow cells. (used with permission, Ref. 37.)

An alternative type of flow cell was utilized by Sabo et al.[35] to conduct both normal-phase and reverse-phase HPLC/FT-IR experiments. The cell (known as the "Circle" cell™) is configured with a cylindrical ZnSe crystal, around which the column effluent can flow. The cell has a 24-µL volume, and ZnSe is compatible with virtually all common HPLC solvents. The infrared beam enters the crystal at

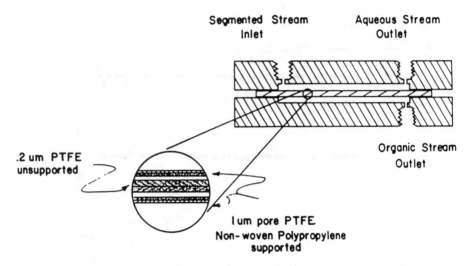

Figure 7.5. Cross-sectional view of the membrane phase separator for reverse-phase HPLC/FT-IR. (Used with permission, Ref. 34.)

one end, reflects off the internal surfaces of the crystal, then exits the crystal at the other end. If a material along the outer surface of the crystal absorbs infrared radiation, a comparison of the light entering and exiting the crystal exhibits the absorption spectrum of the material. This process is referred to as *attenuated total reflectance* (ATR) and is commonly used for obtaining spectra of opaque samples. As we have indicated, for extended path-length transmission cells, water is essentially opaque, so the Circle cell can be used with aqueous mobile phases. The Circle cell, however, suffers from a lack of sensitivity; with solvent subtraction, detection limits of the order of 1–2% (w/v) were reported.[35]

These examples indicate applications and limitations of flow-cell HPLC/FT-IR using large-bore (nominally ≥ 4 mm) columns. The solvent can be a serious interference in the infrared spectra of solutes; so path lengths must be chosen carefully. Furthermore, the detection limits are of the order of 1 μg in reverse phase for compounds with strong, isolated infrared peaks and only a little lower for normal phase. This value is not a serious limitation for gel permeation/size exclusion chromatography applications, where quantities of solute larger than 1 μg do not necessarily degrade column performance, and 1 μg is a satisfactory level for some HPLC applications, as has been indicated. For improved detection limits in flow-cell HPLC/FT-IR, it is necessary to use microbore columns.

7.2.2. Microbore-Column HPLC

One of the difficulties in using FT-IR flow cells with analytical HPLC columns is that the concentration of solute in the flow cell at any given time is quite low. Microbore columns with their lower flow rates and reduced solvent consumption offer the possibility of having higher concentrations in a flow cell; therefore, detection by FT-IR can be more sensitive. With the lower solvent use, it is also economically feasible to use more expensive deuterated solvents whose absorptions, as indicated in the previous section, do not overlap as severely as those of the solute.[36]

In some early work, Brown and Taylor[37] used a microbore polar amino cyano (PAC) column with $CDCl_3$ as the mobile phase to study some of the same process solvents and model amine compounds that had been analyzed previously using analytical columns with the same stationary phase.[29] Amateis and Taylor used microbore amino (NH_2) bonded phases with mobile phase consisting of 0.02% triethylamine added to either deuterated chloroform[38] or a 70:30 $CDCl_3:CCl_4$ mixture[39] to study basic nitrogen compounds (azaarenes and aromatic amines) in coal-derived liquids.

With the commercially available cell used in these studies (0.2-mm path length, 3.2 μL volume), detection limits were improved to 600–700 ng for the test compound 2,6-di-t-butylphenol. For other compounds, detection limits were as large as 1.5 μg with optimized chromatographic conditions and as large as 2.5 μg with column overloading.[40] However, detection limits for ester and amide derivatives, which have intense carbonyl absorptions, were in the range of 200–500 ng.

A major improvement in detection limits for normal-phase flow-cell HPLC/FT-IR was made with a specially constructed "zero-dead-volume" (ZDV) cell.[41] Calcium fluoride (CaF_2) and potassium bromide (KBr) cells were constructed by drill-

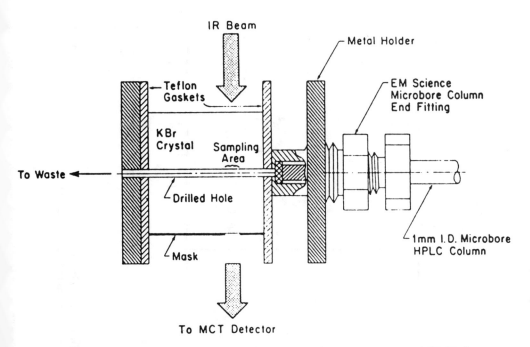

Figure 7.6. Cross-sectional view of the zero-dead-volume microbore HPLC/FT-IR flow cell. (Used with permission, Ref. 41.)

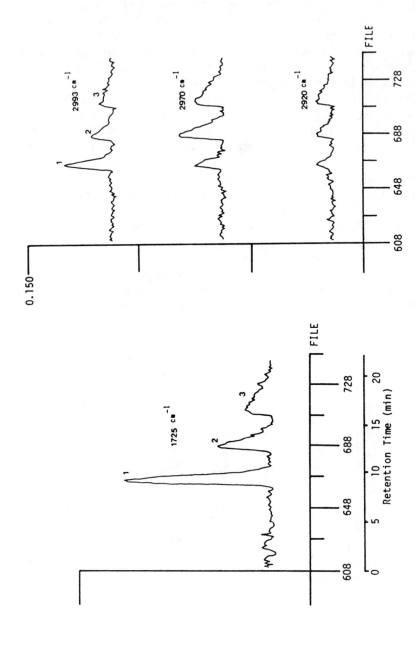

Figure 7.7. Infrared chromatograms for phthalates at four different frequencies. Peak assignments: (1) diethylphthalate, (2) di-*n*-butylphthalate, and (3) di-*n*-pentylphthalate. (Used with permission, Ref. 43.)

ing a 0.75-mm hole into a block of the window material. The effluent of a 1-mm microbore column was directed into the hole, and the infrared beam intersected the effluent flow at an angle of 90°. (See Figure 7.6.) To match the size of the sampling area optically, the 3-mm infrared beam was sent through a 4× beam condenser before it reached the flow cell. These design factors resulted in a flow cell with an effective path length of 0.45 mm and a volume of 0.33 μL; the detection limit for 2,6-di-t-butylphenol was determined to be in the range of 40–50 ng.

Reverse-phase HPLC/FT-IR using microbore columns suffers from the need to use aqueous solvents, but deuterated materials provide some of the same advantages as indicated for normal-phase HPLC. Jinno and Fujimoto demonstrated the feasibility of using PTFE as a flow-cell material for reverse-phase HPLC/FT-IR by flattening a piece of PTFE tubing and affixing it to the exit of an unpacked section of 350-μm i.d. fused silica.[42] With deuterium oxide as the mobile phase, methanol was detected in the system. The same PTFE cell was also used to detect n-hexane eluting from a silica-packed column with N-trifluoromethylperfluormorpholine as the mobile phase.

In followup work, Jinno et al. used a C_{18} packing material in a 0.5-mm i.d. PTFE column and fashioned a flow cell by flattening the end of the column.[43] The optical path length was approximately 30 μm, as in their previous work, and D_2O/CD_3CN was used as the mobile phase. This solvent/cell system allowed useful data to be obtained over the frequency ranges 4000–2600, 2250–1900, and 1750–1600 cm^{-1}. These are the only useful ranges because of absorptions of D_2O, CD_3CN, and PTFE. Various phthalates were used to demonstrate that carbonyl and C–H stretching absorptions could be monitored (see Figure 7.7), but the detection limits were estimated to be only 1 and 10 μg, respectively, for those two frequency regions.

A flow cell constructed using AgCl windows (0.4 mm path length) was used for reverse-phase separations of phthalates using H_2O/CH_3CN and D_2O/CD_3CN mobile phases (90:10), normal-phase separations of cholestanones and cholesterol using a $C_6D_{12}/CDCl_3/CD_3OD$ mobile phase (90.3:9.1:0.6), and size-exclusion chromatography separations of alcohols using octadeuteriotoluene.[44] The AgCl windows opened regions where the PTFE cell of earlier design absorbed the infrared radiation; however, with a 0.4-mm path length, large fractions of D_2O in the mobile phase could not be used.

The application of size-exclusion chromatography with flow-cell FT-IR detection to proteins and other biologically important materials were reported recently.[45] Buffers in deuterated water were used in the mobile phase, and the possibility of studying dynamic properties such as protein folding in solution was discussed.

By replacing analytical columns with microbore columns and redesigning flow cells, it was possible to improve detection limits substantially for normal-phase HPLC/FT-IR but not significantly for reverse-phase HPLC/FT-IR. The fundamental problem with the latter is the limitation in path length imposed by the strong absorptions of water and other reverse-phase solvents. It is this limitation more than any other that drives the developments in solvent-elimination HPLC/FT-IR discussed in the next sections.

7.3. Solvent-Elimination Techniques for HPLC/FT-IR

7.3.1. Analytical-Column HPLC

Relative to flow-cell interfaces, solvent-elimination systems for HPLC/FT-IR can be used to concentrate analytes before spectral data collection and to obtain complete infrared spectra free of spectral interferences due to the mobile phase. With analytical columns, minor components in the analyte mixture can be very dilute, and the amount of solvent that must be removed can be substantial with flow rates typically of the order of 1 mL/min. As we shall see, the potential advantages of solvent-elimination systems can only be gained at the cost of simplicity.

The first solvent-elimination HPLC/FT-IR interface for analytical columns was proposed by Griffiths.[46] This system included four relatively short lightpipes, similar to those used for GC/FT-IR, and HPLC effluent was sent into the lightpipes, where the solvent evaporated. The lightpipes were arranged in a carousel that rotated 90° for each of the four steps in the process—sample deposition, infrared measurement, washing, and drying. The process could be nearly continuous as long as each step proceeded properly, but uneven deposition, incomplete washing, and incomplete drying were among the problems encountered.

With these problems in mind, Kuehl and Griffiths[47,48] designed a substantially different interface. It consisted of a concentrator tube through which the HPLC effluent passed on its way to a carousel of cups filled with KCl powder. The KCl powder was used as a substrate for infrared analysis by diffuse reflectance (DRIFT) spectroscopy. The glass concentrator tube was wrapped with nichrome wire so that heating could assist in the evaporation of the solvent. At the top of the concentrator, near where the HPLC effluent entered, was a small tube through which a stream of air was directed down the tube to enhance the evaporation process (see Figure 7.8). The fate of solvent droplets that traveled to the bottom of the tube was determined by a conventional UV detector in that the droplets containing UV-absorbing materials were allowed to proceed onto the KCl powder while all other droplets were aspirated to waste. A computer-controlled solenoid valve on the aspirator line at the bottom of the concentrator was triggered by the UV detector after a sufficient time delay to ensure that the solvent had traveled from the UV detector to the concentrator. This process minimized the number of sample cups used during a chromatographic experiment, but compounds without UV absorbances were not analyzed by the infrared spectrometer.

Any residual solvent remaining with the analyte in the sample cups was removed by passing air or nitrogen through the cup. For normal-phase solvents, the system operated very efficiently, and the diffuse reflectance technique was sensitive enough to provide detection limits in the submicrogram region for many of the compounds studied. An example of 1 μg of "butter yellow" that eluted in a mobile phase consisting of 2% methanol in n-hexane is shown in Figure 7.9. A limitation of the system is that reverse-phase solvents cannot be eliminated quickly enough, and, of course, any residual water that might remain would adversely affect the KCl substrate.

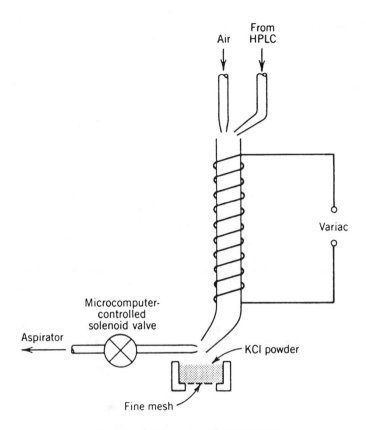

Figure 7.8. Concentrator for solvent-elimination HPLC/FT-IR for analytical columns and diffuse reflectance infrared analysis. (Used with permission, Ref. 47.)

A modification to this basic system allowed it to be used for aqueous reverse-phase HPLC.[49] Shown schematically in Figure 7.10 are the additional items that were placed between the effluent of the HPLC and the concentrator. The system is designed to mix methylene chloride with the HPLC effluent, send the two phases through an extraction coil, and separate the phases. The aqueous phase, which is less dense than the organic layer, was removed by aspirating it off after the flowing two-phase system reverted to a laminar flow at some point beyond the extraction coil. The remaining methylene chloride fraction was directed to the concentrator and DRIFT accessory. Detection limits were in the submicrogram range, and high-quality spectra of "butter yellow" were obtained for 250 ng of sample injected into a C_{18} column using an 80:20 methanol/water mobile phase. Essentially, the detection limits of the reverse-phase system were not significantly different from those obtained with the normal-phase version of the interface.

This interface has some interesting features that are worth nothing. For example,

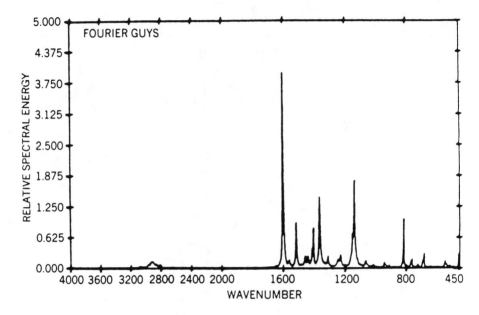

Figure 7.9. Diffuse reflectance spectrum of 1 μg (injected) of "butter yellow" dye obtained in a normal-phase HPLC/FT-IR experiment. (Used with permission, Ref. 47.)

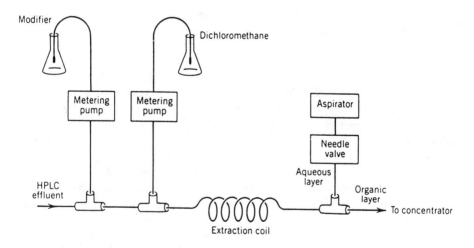

Figure 7.10. Diagram of the additional components required to accommodate aqueous reverse-phase separations prior to sending effluent to the concentrator shown in Figure 7.8. (Used with permission, Ref. 49.)

there is the option to add modifiers (postcolumn) to the aqueous phase to enhance the extraction process or to extract certain classes of compounds selectively. Additionally, buffers used in the mobile phase would certainly remain in the aqueous phase and, therefore, not be deposited on the diffuse reflectance substrate. A potential limitation of the system is the small but finite solubility of water in methylene chloride. Also, if organic modifiers are used in the chromatographic solvent, they may affect the efficiency of the extraction and phase-separation processes.

Another proposed reverse-phase HPLC/FT-IR interface is modeled after the thermospray devices used in HPLC/MS. Griffiths and Conroy described gratifying preliminary results if the spray was directed onto diamond dust as the diffuse reflectance substrate.[14] Details of their interface have not been reported, but, in general, the effluent of the HPLC was heated to 140–150°C and allowed to expand into a vacuum where it was sprayed onto the diamond dust. Microgram quantities were required for satisfactory spectra, but the interface was able to remove a volatile ion-suppression agent like acetic acid. Additional work with this interface will be summarized in the next section.

An HPLC/FT-IR interface in which the water was removed by a postcolumn reaction prior to infrared analysis by diffuse reflectance has also been described.[50,51] In particular, 2,2-dimethoxypropane (DMP) reacts with water to produce acetone and methanol, compounds that are sufficiently volatile to be removed without great difficulty:

$$(CH_3)_2C(OCH_3)_2 + H_2O \longrightarrow O = C (CH_3)_2 + 2 CH_3OH.$$

For analytical columns, the flow rates are sufficient that the amount of solution exiting the postcolumn reactor far exceeds the capacity of the diffuse reflectance accessory; so either of two approaches must be taken to accommodate the flow. A concentrator can be used to remove the bulk of the excess solvent, or a splitter can be installed to direct a small portion of the flow to the infrared cell (see Figure 7.11). The various concentrators serve the same purpose as that described by Kuehl and Griffiths,[47,48] but those shown in Figure 7.11 use aspiration to assist the removal of solvent vapors. In one design the effluent of the concentrator was allowed to drop onto the KCl surface, whereas the other two interaces in the figure indicate the use of a nebulizer for deposition. Nebulization assists evaporation of residual solvent during the deposition of solute onto the DRIFT substrate.

In these interfaces, the KCl substrate was supported in a series of compartments machined into a flat aluminum rod. This so-called "train"[52] was translated past the effluent of the concentrator or nebulizer and then into a diffuse reflectance accessory mounted in the FT-IR spectrometer. In order to minimize distortion of the KCl surface, the KCl was loaded into the train as a slurry and allowed to dry.

Detection limits for the system were of the order of 1 µg for solvents with relatively low levels of water and were as high as 3 µg for some compounds with large amounts of water ($\geq$ 70%) in the mobile phase. The difference rests in the requirement of larger amounts of DMP for the latter solvents and the accompanying dilution of the analytes. On the other hand, the system is capable of performing

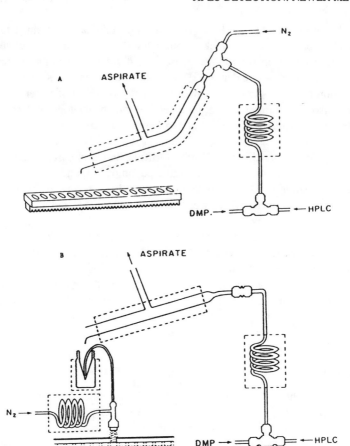

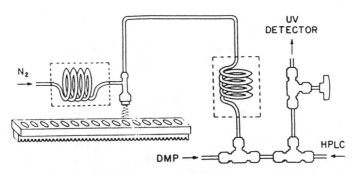

Figure 7.11. Diagram of concentrator- and splitter-based HPLC/FT-IR interfaces that use the 2,2-dimethoxypropane (DMP) postcolumn reaction for aqueous reverse-phase solvents. (Used with permission, Ref. 51.)

normal-phase HPLC/FT-IR experiments in essentially the same manner as the interface of Kuehl and Griffiths[47,48] by simply turning off the postcolumn DMP pump.

As in the case of flow-cell HPLC/FT-IR interfaces, the large quantity of solvent is a limiting factor for solvent-elimination HPLC/FT-IR. Consequently, as we shall see in the next section, there are distinct advantages in working with microbore columns.

7.3.2. Microbore-Column HPLC

Microbore HPLC/FT-IR interfaces were reviewed by Jinno in 1989,[19] so we shall try to highlight the solvent-elimination systems covered in that article and focus our attention on the more recent developments. Since this type of interface probably represents the future of HPLC/FT-IR, it is worth noting that the detection limits are indeed better than those for analytical columns for both normal- and reverse-phase experiments. Needless to say, the normal-phase interfaces were developed first, and we shall begin this section by considering them.

A short communication by Jinno and Fujimoto[53] appears to be the first report of solvent-elimination microbore HPLC/FT-IR, and it was followed very shortly thereafter by a second article.[54] A 0.5-mm i.d. PTFE column packed with silica was used to study a test mixture consisting of three 2,4-dinitrophenylhydrazones. With the n-hexane/methylene chloride mobile phase (65:35) flowing at the rate of 5 μL/min, it was relatively easy to eliminate the solvent as it was deposited in a continuous band onto a KBr plate (8.8×35 mm^2; 5 mm thick). The KBr plate was mounted as shown in Figure 7.12 and translated under a stainless steel wire, down which the HPLC effluent flowed. Infrared spectra were obtained by subjecting the KBr to a transmission experiment with a $3\times$ beam condenser. Detection limits were less than 1 μg, but no spectra were shown in the initial articles. Subsequent work with the interface involved the analysis of metal complexes of the hydrazones[55] using normal-phase HPLC and size-exclusion chromatography of polystyrenes using tetrahydrofuran as the mobile phase.[56]

A modification of this basic interface involved replacing the KBr plate in a linear translation stage with a KBr disk that rotated as the HPLC effluent was deposited.[57] A schematic is shown in Figure 7.13, and it was anticipated that effluents from a 3-h chromatogram could be held on one disk if necessary. It was suggested that the detection limits were less than 100 ng,[19] but 6 μg of each of three cholesterol derivatives separated on a silica column with a cyclohexane/chloroform/methanol mobile phase were used to demonstrate the system in a 2-h chromatogram (see Figure 7.14). It is reported that the interface is now commercially available.

For reverse-phase HPLC, Jinno and co-workers replaced the KBr plate of their original solvent-elimination microbore interface with stainless steel wire nets (SSWN).[58] Of the various mesh sizes evaluated, the screens with 15-μm openings were found to give the best spectra. Detection limits were of the order of 500 ng for carbaryl.

A second type of normal-phase microbore HPLC/FT-IR interface is a compact version of that described by Kuehl and Griffiths.[47,48] Conroy et al.[59] constructed a

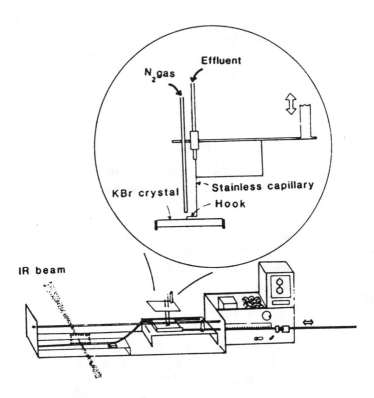

Figure 7.12. Diagram of the microbore HPLC/IR device in which normal-phase effluent is deposited onto a KBr plate. (Used with permission, Ref. 54.)

carousel with a diameter of 120 mm, which contained 180 compartments, into which solutes were deposited. Diffuse reflectance spectrometry from a KCl substrate was used to collect infrared spectra. The sensitivity of diffuse reflectance relative to simple transmission spectroscopy is apparent from the detection limits, which were 200–300 ng for poor samples. Spectra of 10 ng of 4-chloronitrobenzene were recorded after the material eluted from a 1-mm i.d. column in a mobile phase of 2% methanol in hexane.

Another diffuse-reflectance-based interface that uses the postcolumn 2,2-dimethoxypropane (DMP) reaction described in the previous section has been applied to normal- and reverse-phase microbore separations.[60] The effluent was sprayed onto a segmented "train" or continuous trough of KCl either through a simple low-dead-volume "tee" or with a nebulizer. Figure 7.15 shows a spectrum that was collected after a 1-μg injection, but the distribution of the sample over different compartments of the train suggested that the spectrum is of approximately 100 ng. Conservatively, detection limits are in the submicrogram range, but a clear demonstration of this would require an improved nebulizer such as those to be described. Griffiths and Conroy[14] conducted some preliminary work with the DMP reaction

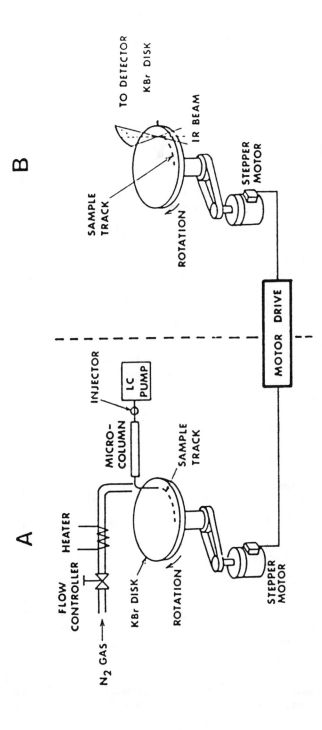

Figure 7.13. Diagram of the microbore HPLC/IR device in which effluent is deposited onto a KBr disk. (A: Effluent is deposited on KBr disk, B: Infrared spectra are recorded) (Used with permission, Ref. 57.)

147

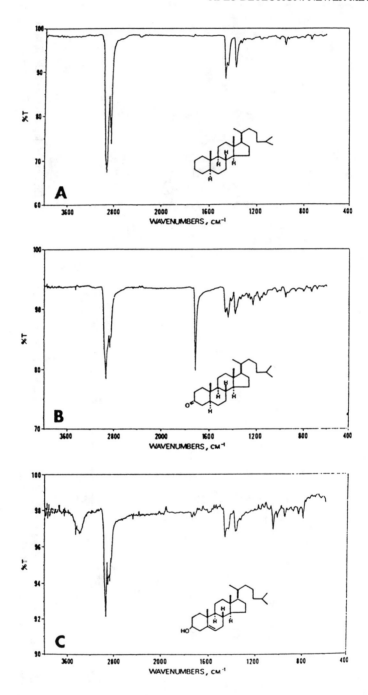

Figure 7.14. Infrared spectra of (a) 5-cholestane, (b) 5-cholestan-3-one, and (c) cholesterol recorded with the interface shown in Figure 7.13. (Used with permission, Ref. 57.)

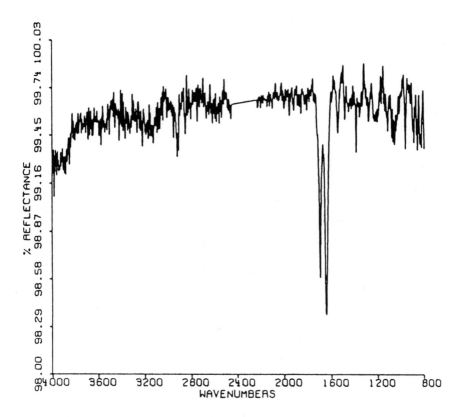

Figure 7.15.   Infrared spectrum of a portion of a 1-μg injection of caffeine following the DMP reaction to remove water from the mobile phase. (Used with permission, Ref. 60.)

and their own deposition system. Using their more sensitive DRIFT accessory, they estimated the detection limits to be approximately 100 ng (injected).

Gradient-elution HPLC/FT-IR with this system has been described, and various improvements in the design have been considered.[61] Of particular interest is a bonded-phase postcolumn reactor, which would eliminate the need for an additional pump to supply the DMP for the water removal step. Specifically, multifunctional compounds like 3,3-dimethoxypropyltrimethoxysilane have been bonded to silica particles and packed into columns.[62] When water comes into contact with the dimethoxypropyl group, an aldehyde and methanol are formed. The aldehyde is part of the bonded phase, but the newly formed methanol elutes with the remaining mobile phase. The amount of solvent to be eliminated is reduced, and the reaction column can be regenerated after the chromatography experiments are concluded. The reaction column does introduce some dead volume, but its effects on the chromatographic separations can be minimized.

Castles et al. described the interface shown in Figure 7.16, in which reverse-phase solvents were eliminated directly with the aid of nebulization for sample

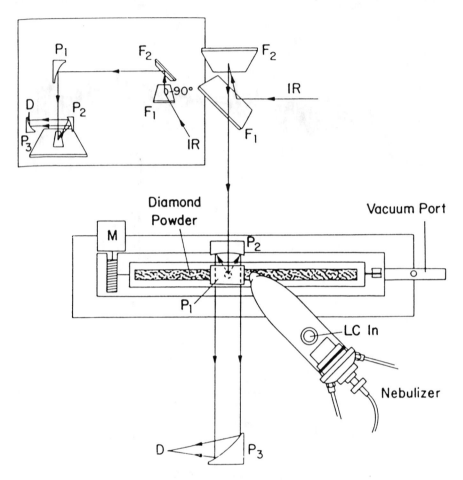

Figure 7.16. Diagram of the reverse-phase HPLC/FT-IR interface, which utilizes an ultrasonic nebulizer for sample deposition and diamond powder for the diffuse reflectance substrate. (Used with permission, Ref. 63.)

deposition and a vacuum port situated near the diffuse reflectance substrate.[63] An ultrasonic nebulizer with a piezoelectric transducer was used to induce desolvation and deposit the sample. The intent was to eliminate all traces of solvent by the time the sample reached the diffuse reflectance substrate, but there were occasions when this goal was not achieved. With residual water a constant possibility, Castles et al. used diamond powder (600 mesh, 30 μm) instead of the more widely used powdered KCl as the DRIFT substrate. The trough in which the diamond powder was loaded was translated past the nebulizer and then through the optics of the DRIFT accessory. Spectra of high quality were obtained for 3-μg injections of *m*-nitroaniline and 2-methyl-3-nitrophenol, and detection limits are probably of the order of 1 μg.

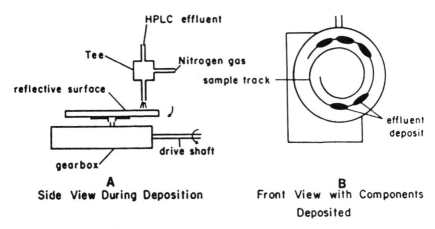

HPLC effluent

Tee

Nitrogen gas

sample track

reflective surface

effluent
deposit

drive shaft

gearbox

A
Side View During Deposition

B
Front View with Components
Deposited

Figure 7.17.   Diagrams of the HPLC/FT-IR interface in which effluent is deposited on a reflective disk and the spiral track formed by the effluent. (Used with permission, Ref. 64.)

Aside from its utility in HPLC/FT-IR experiments, this interface is very important to the development of the field because it represents the first instance where aqueous reverse-phase solvents were eliminated directly, without, for example, extractions or postcolumn reactions. As we shall see, the interfaces that show the most promise are those that use direct elimination of water.

Griffiths and Conroy also used their thermospray apparatus[14] with microbore HPLC and deposited the effluent onto diamond powder. With a mobile phase of 15% methanol in water and a 1-mm i.d. C_{18} column, they separated and obtained identifiable spectra of 20 ng of p-nitrophenol and 2,4-dinitrophenol.

Gagel and Biemann described an apparatus that employed reflection–absorption (R/A) as a means of recording infrared spectra for normal- and reverse-phase HPLC.[64,65] The effluent was deposited onto a circular mirror in a continuous spiral pattern, as shown in Figure 7.17. Direct evaporation of the mobile phase is the fundamental premise of this solvent-elimination scheme, and it is aided by the flow of heated nitrogen gas in the nebulizer assembly (see Figure 7.18). As indicated, the elimination of normal- and reverse-phase solvents (with as much as 55% water) has been demonstrated, and for gradient elution, the temperature of the heated gas stream was varied accordingly to match the solvent composition. Adequate sensitivity was observed with the R/A spectra; Figure 7.19 shows spectra of injected quantities of phenanthrenequinone varying from 248 to 16 ng. While the 16-ng injection was detectable, the spectra were identifiable only with $\geq$ 31 ng of material. A version of this interface that gives a circular pattern for the deposited material is now available commercially.

Griffiths and co-workers have configured an interface that uses a concentric flow nebulizer to deposit the analytes and assist evaporation of the solvent in a vacuum chamber.[66,67] Figure 7.20 shows three nebulizer designs and indicates the types of flow patterns expected at the exits. The nebulizer fits into an evacuable chamber, as

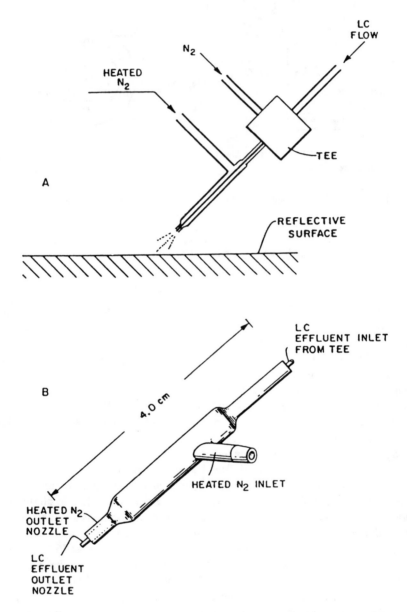

Figure 7.18. Diagrams of the nebulizer used to deposit samples onto the interface in Figure 7.15. (Used with permission, Ref. 65.)

shown in Figure 7.21, and the analytes are deposited on an appropriate substrate. For optimum detection using microscope optics, a spot of analyte should be no larger than approximately 100 μm, and concentric flow nebulization with a sheath of warm helium gas has been used to study normal-[66] and reverse-phase[67] separations with gratifying preliminary results.

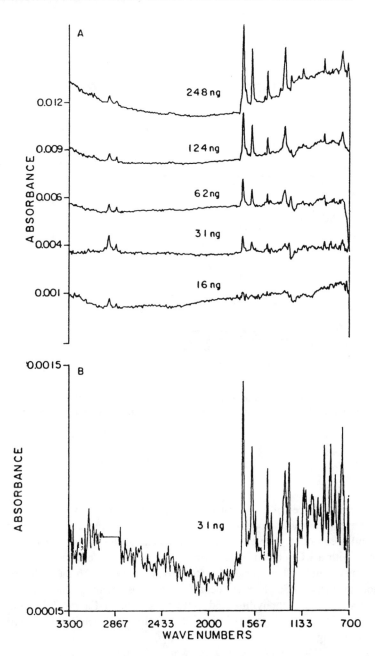

Figure 7.19.   Infrared spectra of phenanthrenequinone for five different deposited quantities using the interface and nebulizer in Figures 7.15 and 7.16. (Used with permission, Ref. 65.)

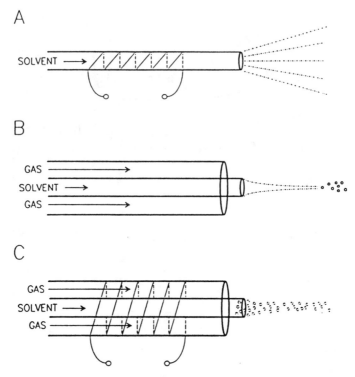

Figure 7.20. Diagram of concentric flow nebulizers: (a) thermospraying, (b) hydrodynamic focusing, and (c) concentric flow nebulization. (Used with permission, Ref. 67.)

For normal-phase solvents, transmission spectra, reflection–absorption spectra, and diffuse reflectance spectra of 50 ng of phenanthrenequinone and 5 ng of acenaphthenequinone were compared. In both cases, the diffuse reflectance spectra obtained with a KCl-coated metal strip as the substrate were superior. In reverse-phase experiments, where residual water might interfere with the KCl strip, transmission through a ZnSe plate was deemed to be most feasible. The spectra of 60 ng each (injected) of theophylline and caffeine shown in Figure 7.22 indicate the sensitivity of the system, and estimates of the minimum identifiable quantity are in the 500 pg range.

One other HPLC/FT-IR interface that warrants some comment is an adaptation of the monodisperse aerosol generation interface for combining liquid chromatography with mass spectrometry (MAGIC-LC/MS).[12] In this system, the HPLC effluent at flow rates up to 0.5 mL/min was sent through a small-diameter glass orifice in order to form a liquid jet. The jet is intersected at 90° by a dispersion gas, thus forming an aerosol. The aerosol is directed into a desolvation chamber and past a skimmer

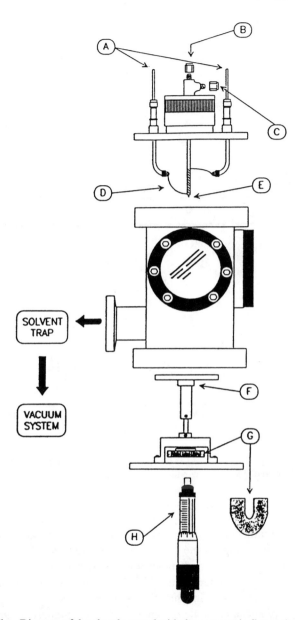

Figure 7.21.   Diagram of the chamber used with the concentric flow nebulizer HPLC/FT-IR interface: A, electrical feedthroughs; B, inlet for chromatographic effluent; C, helium gas inlet; D, heating wire; E, concentric tubes; F, rotatable solute stage; G, magnets for stage rotation; H, micrometer for vertical stage adjustment. (Used with permission, Ref. 67.)

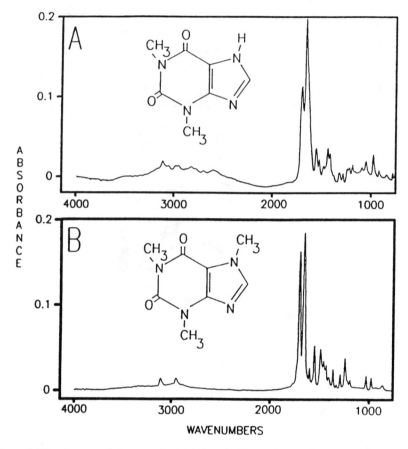

Figure 7.22. Spectra of 60 ng each of (a) theophylline and (b) caffeine using the interface shown in Figures 7.20 and 7.21. (Used with permission, Ref. 67.)

system. The resulting uniform droplets were directed into the ionization chamber of a mass spectrometer. For coupling the MAGIC interface to infrared spectrometry, a KBr plate is placed in the vacuum chamber for sample deposition, then removed and placed into the sample chamber of an FT-IR spectrometer fitted with condensing optics. Preliminary results indicating that this interface is capable of removing aqueous solvents have appeared in the literature,[68,69] but the reported sample quantities are well above the detection limits for the other solvent-elimination reverse-phase HPLC/FT-IR interfaces described previously. It is also not clear yet whether this system will be applied to microbore HPLC only or to analytical columns. Additional work continues at the present time, and it is expected that answers to the various questions regarding this potentially useful mechanism for HPLC/FT-IR will be forthcoming.

7.4. Related Chromatography-Infrared Interfaces

7.4.1. Supercritical Fluid Chromatography

SFC/FT-IR was first predicted as a viable technique for analyzing nonvolatile mixtures by Novotny et al. in 1981.[70] By 1983, Shafer and Griffiths had published preliminary data demonstrating the technique.[17] The interest in using SFC as a hyphenated technique is the compatibility of the mobile phases with infrared spectroscopy. Typical mobile phases used in SFC absorb minimally in the midinfrared region, and the most commonly used solvent is carbon dioxide. As with HPLC interfaces, both flow-cell and solvent-elimination approaches can be taken when combining SFC with FT-IR spectrometry. High-pressure flow cells are typically 5–10 mm in path length and fitted with ZnSe windows.[72,73] Detection limits have been reported in the 300–500 ng range, but recent developments have shown that detection can be achieved with a flow cell down to 50 ng.[74] Pressure/density programming and the addition of modifiers that aid in elution create some spectral interferences for infrared detection. Alternate mobile phases such as Freons and xenon for the infrared interfaces have been investigated.[75] Analytical applications of SFC/FT-IR using a flow cell have included lipids, carbamates, natural products, and propellants.[76–78]

Two approaches have been taken with solvent-elimination SFC/FT-IR. In the first method, infrared analysis using diffuse reflectance spectroscopy was performed after eluents were deposited onto a KCl surface that had been prepared from a slurry.[79] This method has a detection limit of 50 ng. In the second method eluents were deposited onto an infrared transparent window (ZnSe or KBr) and subsequently analyzed by infrared microscopy.[80] This technique has a reported detection level of 30 ng. Applications have also included polymer additives.[81]

7.4.2. Thin-Layer Chromatography

TLC/FT-IR was first performed by measuring the infrared spectra from the TLC plate directly, with the best results employing diffuse reflectance measurements.[82] Detection limits of 1 μg can be obtained, but the adsorbent can exhibit strong interferences in the infrared spectra. Photoacoustic infrared detection of TLC spots has also been demonstrated, and adsorbent interferences were thereby minimized.[83] The bulk of the recent work in the area of TLC/FT-IR has been in transfer of separated zones from TLC media to infrared substrates for subsequent analysis by infrared spectroscopy. Transfer can be accomplished by several different routes, but the infrared measurement technique is typically diffuse reflectance.[84–86] Detection can be obtained at the 500-ng level.

7.4.3. Near-IR HPLC

The use of near-IR as a detector for HPLC was first reported in the literature in 1986.[87] The initial studies were focused on the need for an HPLC detector for aqueous preparative work and nonchromophore-bearing solutes. More recent studies[88,89] have shown near-IR as a universal HPLC detector to have low drift and noise levels, high sensitivity, fast response, and wide linear dynamic range as advantages over conventional detectors. The potential for this technique is enormous.

7.5. The Current Status of HPLC/FT-IR

As of this writing, it is reported that a normal-phase HPLC/FT-IR interface based on the design of Jinno and co-workers[57] and a reverse-phase system (also capable of normal-phase separations) based on that of Gagel and Biemann[64,65] are available commercially. The improvements in the past 12–15 years have brought HPLC/FT-IR to the stage of viability, and it is likely that other interfaces will also be available in the near future.

The commercial interfaces are solvent-elimination devices, and the reasons for that have been alluded to repeatedly in the previous sections. Flow-cell systems have limitations that require significant interface modifications for minor, "everyday" chromatographic changes, but this is no longer true for solvent-elimination systems. Griffiths has discussed a unified approach to chromatography–FT-IR interfaces in attempting to configure a system capable of accepting effluents from a gas chromatograph, a high-performance liquid chromatograph, or a supercritical fluid chromatography.[17] The various designs for each type of interface have evolved so that this is a real possibility. Just a few years ago, analytical and microbore HPLC/FT-IR interfaces were different, and normal- and reverse-phase interfaces were different. Now that all these HPLC experiments can be accommodated by a single interface, it becomes even more likely that HPLC, GC, and SFC can use the same system for coupling to FT-IR spectrometry. For the necessary sensitivity, simple infrared transmission experiments are not sufficient. Diffuse reflectance, reflection–absorption, and microsampling (or microscope) optics are required. Each of these sampling methods is more efficient in an FT-IR spectrometer than in a dispersive instrument.

There are many factors that impinge on the current state-of-the-art in HPLC/FT-IR, but it appears that they have all been identified and are being refined. It is with this knowledge that we feel that HPLC/FT-IR will take its place alongside GC/FT-IR, GC/MS, and HPLC/MS as a useful, reliable tool in the analytical laboratory.

7.6. References

1. G. Herzberg, *Infrared and Raman Spectra of Polyatomic Molecules*, Van Nostrand Reinhold: New York (1945).

2. N. B. Colthup, L. H. Daly, and S. E. Wiberley, *Introduction to Infrared and Raman Spectroscopy*, Academic: New York (1990).

3. R. M. Silverstein, G. C. Bassler, and T. C. Morrill, *Spectrometric Identification of Organic Compounds*, John Wiley: New York (1981).

4. J. R. Durig, ed., *Vibrational Spectra and Structure,* Vol. 12, Elsevier: Amsterdam (1983).

5. P. R. Griffiths and J. A. deHaseth, *Fourier Transform Infrared Spectrometry,* John Wiley and Sons: New York Ch. 19, 611–46 (1986).

6. R. White, *Chromatography/Fourier Transform Infrared Spectroscopy and Its Applications,* Marcel Dekker: New York Ch. 3, 95–136 (1990).

7. J. F. Holland, C. G. Enke, J. Allison, J. T. Stults, J. D. Pinkston, B. Newcome, and J. T. Watson, *Anal. Chem.* **55** (1983) 997A–1012A.

8. P. R. Griffiths, J. A. deHaseth, and L. V. Azarraga, *Anal. Chem.* **55** (1983) 1361A–1387A.

9. S. Bourne, G. Reedy, P. Coffey, and D. Mattson, *Am. Lab.* (Fairfield, CT) **16**(6) (1984) 90–95.

10. S. Bourne, A. M. Haefner, K. L. Norton, and P. R. Griffiths, *Anal. Chem.* **62** (1990) 2448–52.

11. L. Yang, G. J. Fergusson, and M. L. Vestal, Anal. Chem. (1984), **56,** 2632–2636.

12. R. C. Willoughby and R. F. Browner, *Anal. Chem.* **56** (1984) 2626–31.

13. K. Jinno, in *Detectors for Liquid Chromatography,* E. S. Yeung, ed., John Wiley and Sons: New York, Ch. 3, 64–104 (1986).

14. P. R. Griffiths and C. M. Conroy, *Adv. Chromatogr.* **25** (1986) 105–38.

15. J. W. Hellgeth and L. T. Taylor, *J. Chromatogr. Sci.* **24** (1986) 519–28.

16. P. R. Griffiths, S. L. Pentoney, Jr., A. Giorgetti, and K. H. Shafer, *Anal. Chem.* **58** (1986) 1349A–1366A.

17. P. R. Griffiths, in *Analytical Applications of Spectroscopy,* C. S. Creaser and A. M. C. Davies, eds., Royal Society of Chemistry: London, 173–87 (1988).

18. W. B. Crummett, H. J. Cortes, T. G. Fawcett, G. J. Kallos, S. J. Martin, C. L. Putzig, J. C. Tou, V. T. Turkelson, L. Yurga, and D. Zakett, *Talanta* **36** (1989) 63–87.

19. C. Fujimoto and K. Jinno, *Tr. Anal. Chem.* **8**(3) (1989) 90–96.

20. K. S. Kalasinsky and V. F. Kalasinsky, in *HPLC in the Pharmaceutical Industry,* G. W. Fong and S. K. Lam, eds., Marcel Dekker: New York, 147–69 (1991).

21. W. D. Perkins, *J. Chem. Educ.* **63** (1986) A5–A10.

22. W. D. Perkins, *J. Chem. Educ.* **64** (1987) A269–A271.

23. W. D. Perkins, *J. Chem. Educ.* **64** (1987) A296–A305.

24. D. W. Vidrine, in *Fourier Transform Infrared Spectroscopy—Applications to Chemical Systems,* J. R. Ferraro and L. J. Basile, eds., Academic: New York, Vol. 2, Ch. 4, 129–64 (1979).

25. K. L. Kizer, A. W. Mantz, and L. C. Bonar, *Am. Lab.* (Fairfield, CT) **7**(5) (1975) 85–90.

26. D. W. Vidrine and D. R. Mattson, *Appl. Spectrosc.* **32** (1978) 502–6.

27. D. W. Vidrine, *J. Chromatogr. Sci.* **17** (1979) 477–85.

28. R. S. Brown and L. T. Taylor, *Anal. Chem.* **55** (1983) 723–30.

29. P. G. Amateis and L. T. Taylor, *Chromatographia* **18** (1984) 175–82.

30. S. Wachholz, H. Geisler, and J. Bleck, *J. Liq. Chromatogr.* **11** (1988) 779–92.

31. C. C. Johnson and L. T. Taylor, *Anal. Chem.* **55** (1983) 436–41.

32. C. W. Saunders and L. T. Taylor, *Appl. Spectrosc.* **45** (1991) 900–5.

33. C. C. Johnson, J. W. Hellgeth, and L. T. Taylor, *Anal. Chem.* **57** (1985) 610–15.

34. J. W. Hellgeth and L. T. Taylor, *Anal. Chem.* **59** (1987) 295–300.

35. M. Sabo, J. Gross, J.-S. Wang, and I. E. Rosenberg, *Anal. Chem.* **57** (1985) 1822–26.

36. L. T. Taylor, *J. Chromatogr. Sci.* **23** (1985) 265–72.

37. R. S. Brown and L. T. Taylor, *Anal. Chem.* **55** (1983) 1492–97.

38. P. G. Amateis and L. T. Taylor, *LC* **2**(11) (1984) 854–57.

39. P. G. Amateis and L. T. Taylor, *Anal. Chem.* **56** (1984) 966–71.

40. R. S. Brown, P. G. Amateis, and L. T. Taylor, *Chromatographia* **18**(7) (1984) 396–400.

41. C. C. Johnson and L. T. Taylor, *Anal. Chem.* **56** (1984) 2642–47.

42. K. Jinno and C. Fujimoto, *Chromatographia* **17**(5) (1983) 259–61.

43. K. Jinno, C. Fujimoto, and G. Uematsu, *Am. Lab.* (Fairfield, CT) **16**(2) (1984) 39–45.

44. C. Fujimoto, G. Uematsu, and K. Jinno, *Chromatographia* **20**(2) (1985) 112–16.

45. E. E. Remsen and J. J. Freeman, *Appl. Spectrosc.* **45** (1991) 868–73.

46. P. R. Griffiths, *Appl. Spectrosc.* **31** (1977) 497–505.

47. D. Kuehl and P. R. Griffiths, *J. Chromatogr. Sci.* **17** (1979) 471–78.

48. D. T. Kuehl and P. R. Griffiths, *Anal. Chem.* **52** (1980) 1394–400.

49. C. M. Conroy, P. R. Griffiths, P. J. Duff, and L. V. Azarraga, *Anal. Chem.* **56** (1984) 2636–42.

50. K. S. Kalasinsky and V. F. Kalasinsky, Thirty-Fourth Pittsburgh Conference on Analytical Chemistry and Applied Spectroscopy, March 7–12, 1983, Atlantic City, New Jersey, Paper 357.

51. V. F. Kalasinsky, K. G. Whitehead, R. C. Kenton, J. A. S. Smith, and K. S. Kalasinsky, *J. Chromatogr. Sci.* **25** (1987) 273–80.

52. V. F. Kalasinsky, J. A. S. Smith, and K. S. Kalasinsky, *Appl. Spectrosc.* **39** (1985) 552–54.

53. K. Jinno and Ch. Fujimoto, *J. High Resolut. Chromatogr. Comm.* **4** (1981) 532–33.

54. K. Jinno, C. Fujimoto, and Y. Hirata, *Appl. Spectrosc.* **36** (1982) 67–69.

55. C. Fujimoto, K. Jinno, and Y. Hirata, *J. Chromatogr.* **258** (1983) 81–92.

56. K. Jinno, C. Fujimoto, and D. Ishii, *J. Chromatogr.* **239** (1982) 625–32.

57. C. Fujimoto, T. Morita, K. Jinno, and S. Ochiai, *Chromatographia* **23**(7) (1987) 512–16.

58. C. Fujimoto, T. Oosuka, and K. Jinno, *Anal. Chim. Acta* **178** (1985) 159–67.

59. C. M. Conroy, P. R. Griffiths, and K. Jinno, *Anal. Chem.* **57** (1985) 822–25.

60. K. S. Kalasinsky, J. A. S. Smith, and V. F. Kalasinsky, *Anal. Chem.* **57** (1985) 1969–74.

61. V. F. Kalasinsky, T. H. Pai, P. C. Kenton, and K. S. Kalasinsky, Seventh International Conference on Fourier Transform Spectroscopy, June 19–23, 1989, Fairfax, Virginia, Paper P5.63.

62. T. H. Pai, Ph.D. Dissertation, Mississippi State University (1989).

63. M. A. Castles, L. V. Azarraga, and L. A. Carreira, *Appl. Spectrosc.* **40** (1986) 673–80.

64. J. J. Gagel and K. Biemann, *Anal. Chem.* **58** (1986) 2184–89.

65. J. J. Gagel and K. Biemann, *Anal. Chem.* **59** (1987) 1266–72.

66. D. J. J. Fraser, K. L. Norton, and P. R. Griffiths, in *Infrared Microspectroscopy, Theory and Applications,* R. G. Messerschmidt and M. A. Harthcock, eds., Marcel Dekker: New York, Ch. 14, 197–210 (1987).

67. A. J. Lange, P. R. Griffiths, and D. J. J. Fraser, *Anal. Chem.* **63** (1991) 782–787.

68. R. M. Robertson, J. A. deHaseth, J. D. Kirk, and R. F. Browner, *Appl. Spectrosc.* **42** (1988) 1365–68.

69. R. M. Robertson, J. A. deHaseth, and R. F. Browner, *Appl. Spectrosc.* **44** (1990) 8–13.

70. M. Novotny, S. R. Springston, P. A. Peaden, J. C. Fjeldstead, and M. L. Lee, *Anal. Chem.* **53** (1981) 407A–414A.

71. K. H. Shafer and P. G. Griffiths, *Anal. Chem.* **55** (1983) 1939–42.

72. C. C. Johnson, J. W. Jordan, and L. T. Taylor, *Chromatographia* **20**(12) (1985) 717–23.

73. J. W. Jordan and L. T. Taylor, *J. Chromatogr. Sci.* **24** (1986) 82–88.

74. R. C. Wieboldt, G. E. Adams, and D. W. Later, *Anal. Chem.* **60** (1988) 2422–27.

75. K. Jinno, *Chromatographia* **23** (1) (1987) 55–62.

76. M. E. Hughes and J. E. Long, *J. Chromatogr. Sci.* **24** (1985) 535–40.

77. R. C. Wieboldt and J. A. Smith, in *Supercritical Fluid Extraction and Chromatography—Techniques and Applications,* B. A. Charpentier and M. R. Sevenants, eds., ACS Symposium Series 366, American Chemical Society: Washington, DC, 229–42 (1988).

78. M. Ashraf-Khorassani and L. T. Taylor, *Anal. Chem.* **61** (1989) 145–48.

79. K. H. Shafer, S. L. Pentoney, Jr., and P. R. Griffiths, *Anal. Chem.* **58** (1986) 58–64.

80. S. L. Pentoney, Jr., K. H. Shafer, and P. R. Griffiths, *J. Chromatogr. Sci.* **24** (1986) 230–35.

81. M. W. Raynor, K. D. Bartle, I. L. Davies, A. Williams, A. A. Clifford, J. M. Chalmers, and B. W. Cook, *Anal. Chem.* **60** (1988) 427–33.

82. M. P. Fuller and P. R. Griffiths, *Anal. Chem.* **50** (1978) 1906–10.

83. R. White, *Anal. Chem.* **57** (1985) 1819–22.

84. K. H. Shafer and P. R. Griffiths, *Anal. Chem.* **58** (1986) 2708–14.

85. K. H. Shafer, J. A. Herman, and H. Bui, *Amer. Lab.* (Fairfield, CT) **20**(2) (1988) 142–47.

86. H. Yamamoto, K. Wada, T. Tajima, and K. Ichimura, *Appl. Spectrosc.* **45**(2) (1991) 253–59.

87. E. M. Ciurczak and M. B. Weis, *Spectrosc.* **2**(10) (1986) 33–38.

88. E. M. Ciurczak and T. A. Dickson, *Spectrosc.* **6**(2) (1991) 12–17.

89. E. M. Ciurczak, private communication.

CHAPTER

8

HPLC Detection by Mass Spectrometry

Kenneth B. Tomer

Laboratory of Molecular Biophysics
National Institute of Environmental Health Sciences
P.O. Box 12233
Research Triangle Park, North Carolina 27709

8.1. Introduction

An ideal detector system for HPLC should combine optimum sensitivity with maximum identification capability. Of all detector systems used with HPLC separations, the mass spectrometer arguably comes closest to being the ideal detector. Sensitivity in the attomole range has been demonstrated under selected ion monitoring conditions, while the structural information, including molecular weight, atomic composition, and constituent identification that can be obtained from low levels of analytes is unsurpassed.

Unfortunately, the difference in the operating environments of HPLC (liquid under pressure) and mass spectrometry (gas phase at high vacuum, ca. 10^{-7} torr) is also the greatest of the various detector systems coupled with HPLC. Although this is a formidable barrier to the coupling of the two techniques, it has also presented an interesting challenge to the scientific community.

In response to this challenge, a variety of ingenious approaches have been devised. None of the approaches tried to date has proved to be perfect—all tend to put constraints on either the HPLC operating conditions, the mass spectrometer's operating conditions, or both. It should be noted that the interaction between chromatographers and mass spectrometrists has not been a one-way street, but that the constraints imposed by the mass spectrometer have driven significant developments in the field of HPLC. In this chapter, the various interfaces that have been devised will be described and compared, with emphasis on the newer techniques.

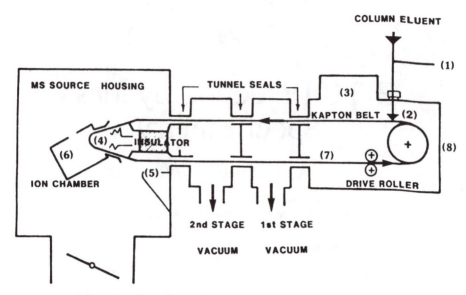

Options: (1) → Drain line (split mode for excess solution removal) or ← post column addition of reagent (FAB mode) (2) Simple contact depositor, frit depositor, jet sprayer (gas nebulized, thermospray). Various angling (45–90°). (3) Solution preheating. Infrared heater in earlier versions. Unnecessary with micro HPLC or FAB. (4) Sample heater. (5) Clean-up heater for EI/CI modes. Not used with FAB. (6) Ionization source (EI, CI, FAB). (7) Belt cleaner (mechanical scrubber, wash bath). (8) Probe aligning mechanism.

Figure 8.1. Schematic of the moving belt interface. (Reprinted from Ref. 4 by permission of John Wiley & Sons, Inc.)

8.2. Interface Designs

8.2.1. Transport Interfaces

Transport interfaces are interface designs in which the LC effluent is transported mechanically into the mass spectrometer's ion source from the end of the HPLC column. The mobile phase is removed prior to entry of the sample into the ion source. The first report of a successful LC/MS interface was a transport interface called the moving wire interface.[1] This interface was not commercialized, however. The first commercially available LC/MS interface, the moving belt, was of the transport type.[2] Although the moving belt interface is still being successfully used to solve problems, it is no longer actively marketed, and the discussion of this interface is primarily included for historical perspective.

A schematic of a generic moving-belt interface is shown in Figure 8.1.[3] The eluent is spray deposited through a heated nebulizer onto the belt. The belt then passes under infrared heaters and through two vacuum locks prior to entering the source. The combination of heaters and vacuum locks effectively removes the mobile phase. The belt passes into the source, and the sample is flash evaporated off the belt. On its return, the belt passes through a scrubber, which removes non-volatile buffers and contaminants.

The belt interface has several advantages over some of the newer interfaces. The belt is compatible with electron impact (EI), chemical ionization (CI), and fast atom bombardment (FAB) ionization modes. It is also compatible with nonvolatile buffers. This latter characteristic is unique among the LC/MS interfaces available today.

The disadvantages of the belt that have helped lead to its effective commercial demise are problems with carryover and its limitation to thermally stabile compounds, that is, compounds whose thermal stabilities lie between those amenable to GC/MS analysis and those amenable only to desorption CI.[4] The integration of the belt with FAB has helped to alleviate some of the thermal instability problem,[5] although problems with precise control of the atom beam flux have been noted.

8.2.2. Direct Liquid Introduction

In the direct liquid introduction (DLI) interface, the column effluent is introduced directly into the instrument. A solvent jet is formed by passing a flow of up to 40 μL/min through a 2–5 μm diameter hole drilled through a replaceable diaphragm at the probe tip (Figure 8.2).[6] If higher mobile phase flow rates are used in the HPLC separation, the excess flow goes into a waste stream located downstream from the orifice, leading to an obvious reduction in the amount of analyte that is presented to the mass spectrometer. For example, for a 1 mL/min flow and a 40 μL/min jet, only 4% of the analyte reaches the ion source. Incorporation of a desolvation chamber permits the use of DLI with conventional CI MS. Water cooling of the interface tip is used to prevent premature evaporation of the solvent.

In addition to restrictions on the flow rate, only volatile buffers can be used with the technique. It is compatible, however, with both normal and reverse-phase systems.

Although DLI interfaces, like the moving belt interface, have been virtually abandoned by the manufacturers, the advent of LC columns with flow rates in the low nanoliter per minute range could revitalize the technique.[7,8]

In our laboratory, we have used wall-coated, chemically bonded, or packed fused-silica or glass capillary LC columns that deliver flow rates of approximately 50 nL/min to the ion source.[9] Due to the nL/min flow rates associated with these columns, we have called them nanoscale capillary LC columns. Wall-coated fused-silica columns have i.d.s of approximately 10 μm and flow rates of 5–50 nL/min, while 50 μm i.d. packed columns also provide flow rates of approximately 50 nL/min. The pumping system of a typical mass spectrometer can handle a flow of about 0.2 μL/min and maintain a source pressure that is compatible with EI spectral acquisition (ca. 5×10^{-6} torr). Thus, direct introduction of the effluent into the ion source is compatible with EI data acquisition.

In our work in this area, we have developed an interface probe for the coated columns in which the column is an integral part of the interface (Figure 8.3). The column is tapered, heated, and inserted directly into the source. Modifications to the

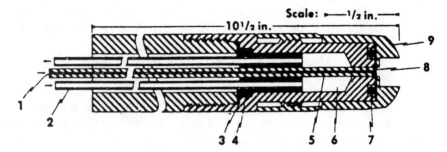

Micro-LC/MS probe interface. (1) Micro-LC effluent line, (2) water-cooling inlet tube, (3) Teflon washer for maintaining vacuum seal between probe tip/cooling chamber and probe shaft, (4) throughput tube collet, (5) 0.004-in. i.d. x 0.062-in. o.d. stainless-steel throughput tube, (6) water cooling chamber, (7) Kalrez O-ring, (8) diaphragm containing 5 -μm pinhole in center, and (9) a removable end cap.

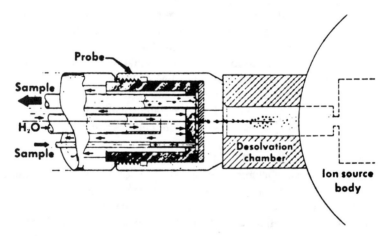

Hewlett-Packard split-effluent DLI LC/MS probe interface.

Figure 8.2. Schematic of the direct liquid introduction interface. (Reprinted with permission from Ref. 6.)

original interface design have been made so that it can be utilized on a magnetic sector instrument. The range of compounds that can be successfully analyzed by this technique appears to be similar to that amenable to analysis by direct insertion probe.

Several other groups have developed similar interfaces in which a length of fused

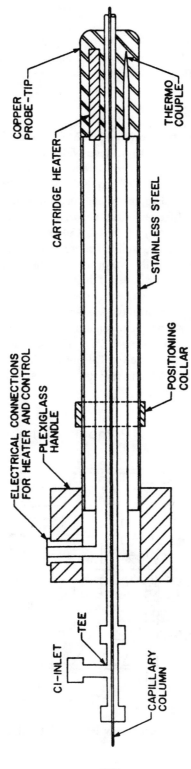

Figure 8.3. Schematic of the open tubular LC interface. (Reprinted with permission from Ref. 9. Copyright 1987 American Chemical Society.)

167

silica is used as a transfer line from a more conventional LC column to the ion source. Reported column flow rates are significantly higher (up to 100 μL/min), and splitting of the LC effluent is usually required.[10–12]

8.2.3. Thermospray

The thermospray (TSP) interface developed by Marvin Vestal[13] is the first commercially available LC/MS interface to achieve widespread acceptance in the mass spectrometry community. The TSP interface requires a dedicated ion source as well as an interface probe and controller (Figure 8.4). In the TSP process, the column effluent flows through a heated vaporizer and forms a supersonic jet. Independent control of the temperature of the vaporizer, vapor, and source block is available and necessary for efficient thermospray operation. Column flow rates of approximately 1 mL/min can be accommodated, with the entire thermospray plume passing into the ion source. Ions are electrically drawn into a sampling cone situated perpendicular to the vapor flow. The remainder of the vapor flow (and sample) is pumped out of the source by means of an additional rough pump.

In the early TSP designs, ionization was achieved through interaction of the analyte with buffer, which is usually ammonium acetate. Thus $(M + H)^+$ and $(M + NH_4)^+$ ions in the positive ion mode and $(M - H)^-$ and $(M + OAc)^-$ ions in the negative ion mode were formed. Newer TSP models include a filament and/or a discharge electrode and a repeller in the source. This removes the necessity of

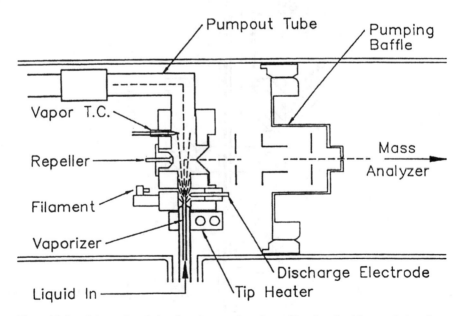

Figure 8.4. Schematic of the thermospray interface. (Reprinted with permission from Vestec Corp.)

having a buffer present for ionization. The extent of fragmentation observed in the mass spectrum can often be altered by changing the repeller voltage (see, e.g., Refs. 14–16).

Generally, TSP/MS is used in conjunction with reverse-phase HPLC. When used with normal-phase LC, the discharge electrode or filament produces ionization in a process that is sometimes referred to as "hot DLI." As with DLI, volatile buffers are preferred. Nonvolatile buffers, such as phosphate or sulfate, tend to plug the vaporizer.

A major advantage that the TSP interface enjoyed over the transport and DLI interfaces was that it proved to be a relatively gentle ionization mode that could be used for relatively nonvolatile and thermally labile compounds, such as phospholipids.[17] This applicability to a wide range of the types of compounds that were typically being separated by HPLC (versus GC) contributed greatly to the rapid acceptance of the technique.

8.2.4. Particle Beam

The particle beam interface was first reported by Willoughby and Browner in 1984 and called the *monodisperse aerosol generation interface* (MAGIC).[18] As the interface became commercialized, the nomenclature rapidly evolved into "particle beam," which better describes the fundamental process.

In this interface (Figure 8.5), the LC effluent passes through a 10 μm i.d. orifice into a desolvation chamber where the liquid breaks up into drops. A perpendicular flow of helium is used to prevent the drops from recoalescing. This process leaves the solute in a dry particulate state that is traveling at a high velocity. The particle beam, solvent vapors, and helium then pass through a momentum separator much like that used for packed column GC/MS. The particle beam at low pressure (10^{-6} torr) enters the ion source, strikes the heated source, and is then vaporized. Compatibility with flow rates of up to 2 mL/min has been reported.[19] More typical flow rates, however, range from 0.1 to 0.5 mL/min.[18] Both organic and aqueous mobile phases are compatible with particle beams, but the sensitivity drops by a factor of 5 or more upon going from 100% methanol to 100% water.[20] Voyksner investigated several solvents and found that sensitivity was inversely proportional to the heat of vaporization of the solvent.[21] Sensitivity has also been observed to be nonlinear at low analyte levels (<5 ng).[22] This has been attributed to a "carrier" effect at higher analyte levels, but is very compound dependent.[23]

Standard EI or CI spectra can be obtained with the technique, and recently integration of particle beam introduction with FAB/MS for the analysis of higher-molecular-weight compounds has been reported.[24]

A variation of the particle beam theme has been developed in which thermospray, membrane separation, and momentum separation are combined. This approach, termed the "Universal Interface," is shown schematically in Figure 8.6.[25] It is claimed to provide excellent performance with mobile phase flow rates as high as 2 mL/min and provides EI spectra or CI spectra with appropriate instrumental operating conditions. Efficient transfer (90%) of analyte from the end of the column to the

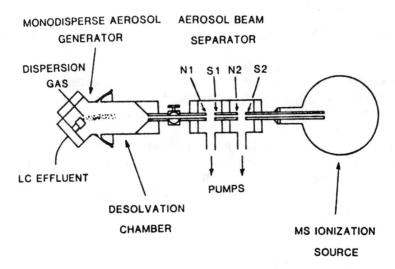

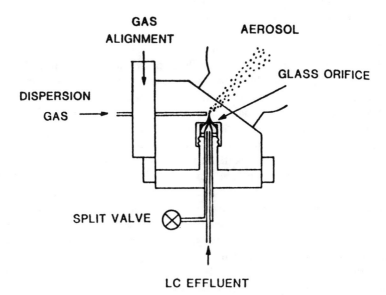

Figure 8.5. Schematic of the particle beam interface and aerosol generator: N1, nozzle 1; N2, nozzle 2; S1, skimmer 1; S2, skimmer 2. (Reprinted with permission from Ref. 18. Copyright 1984 American Chemical Society.)

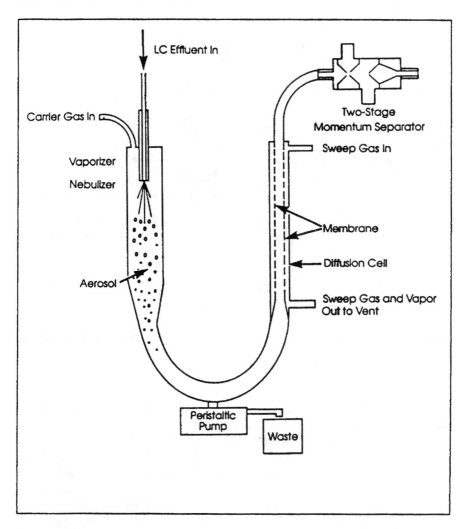

Figure 8.6.   Schematic of the Vestec Universal Interface. (Reprinted with permission from Vestec Corp.)

source is claimed, with enrichment factors of over 10^6 routinely observed. This interface has just been released commercially as of this writing; so very few data have appeared in the literature.

8.2.5. Continuous-Flow Fast Atom Bombardment

The first desorptive ionization process useful for the analysis of large, polar, and/or thermally labile molecules that achieved widespread acceptance was fast atom bombardment (FAB).[26] FAB permits the analysis of a wide range of compounds of

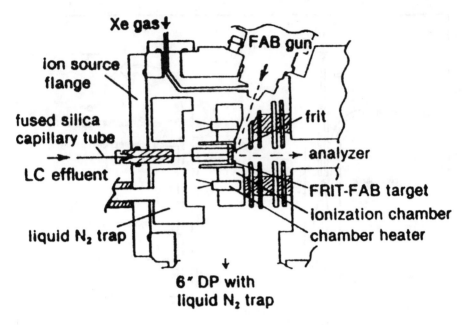

Figure 8.7. Schematic of the Frit FAB interface. (Reprinted with permission from JEOL, Inc.)

interest in biochemistry, such as peptides, nucleotides and oligonucleotides, oligosaccharides, polar conjugates, and bile acids.

Although FAB allowed the analysis of relatively complex mixtures and the determination of analytes that heretofore had been relatively intractable to common MS techniques, there were still some drawbacks to the technique. The presence of a high concentration of liquid matrix (typically glycerol) leads to poor sensitivity, high background, and suppression effects in mixtures. In addition, many chemists had developed elaborate detailed HPLC separation techniques for analytes that were amenable to FAB/MS and desired an on-line method for coupling their HPLC separation with FAB/MS. FAB also is not an ideal technique. To incorporate LC with FAB and to alleviate some of the technique's inherent problems, two similar approaches were developed.

In 1985, Ito and co-workers published a paper on the coupling of micro HPLC with FAB using a length of fused-silica capillary as a transfer line that ended at a stainless-steel frit (Frit FAB) (Figure 8.7).[27] The FAB matrix is incorporated in the mobile phase. Initially, flow rates of ca. 0.5 μL/min were used with the interface. Incorporation of a liquid nitrogen trap in the ion source permits flow rates up to 10 μL/min. For faster LC column flow rates, a splitter is incorporated.

In 1986, Caprioli, Fan, and Cottrell reported on the design of a sample probe that allowed a continuous flow of solution into a FAB source (Figure 8.8)[28,29] known as continuous-flow FAB (CF-FAB).[30] A flow rate of 5 μL/min is used with 5–10% glycerol incorporated in the mobile phase as the FAB matrix. Improved sensitivity

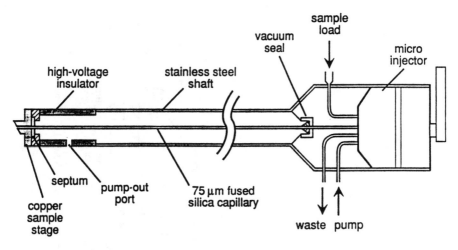

Figure 8.8. Schematic of the ocntinuous-flow FAB interface. (Reprinted with permission from Ref. 28. Copyright 1986 American Chemical Society.)

and decreased ion-suppression effects were observed due to the decreased amount of organic matrix material on the probe tip at any given moment. Improvements to the design have been made since its inception, including a heated FAB source and use of a filter pad to help remove excess liquid. Both of these improvements aid in maintaining a stable source pressure.

Tomer, Jorgenson and co-workers developed a coaxial approach to interfacing nanoscale LC techniques, (Figure 8.9), where the mobile phase flows are approximately 50 nL/min, with CF-FAB.[31,32] In this interface, the analytical column, either a 10 μm i.d., open tube for flow injection analysis or a 50 μm i.d. packed column, is threaded through a larger-bore fused-silica column. The FAB matrix solution is delivered in the intercolumn space. A 10 μm difference in the outer diameter of the analytical column and the inner diameter of the matrix column is typical and results in a matrix flow rate of ca. 0.5 μL/min. This interface design permits independent optimization of both analytical conditions and mobile-phase conditions.

The applicability of the coaxial interface has been extended to 320 μm i.d. columns.[33] With these larger columns, a 25 μm i.d. transfer line serves as the inner column. Again, independent optimization of analytical and matrix flows is achieved. Column flow rates of 3–5 μL/min and matrix flow rates of 1 μL/min are used.

8.2.6. Electrospray/Ion Spray

Electrospray ionization (ESI) and ion spray ionization (ISI) are two inlet/ionization techniques for use with atmospheric pressure ionization (API) (Figure 8.10).[34-37] Because the two approaches to interfacing LC and MS are quite similar

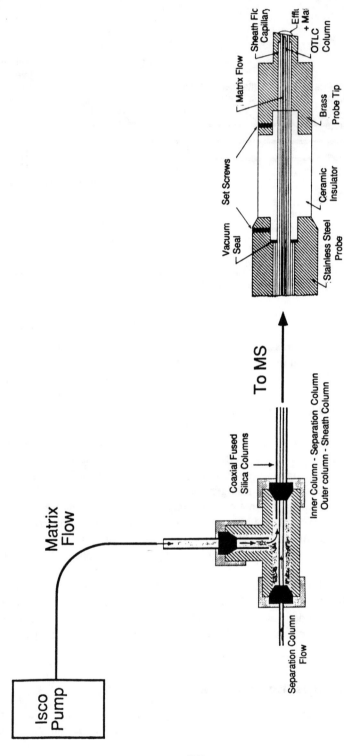

Figure 8.9. Schematic of the coaxial CF-FAB interface. (Reprinted from Ref. 31 by permission of John Wiley & Sons, Ltd. Copyright 1988, John Wiley & Sons, Ltd.)

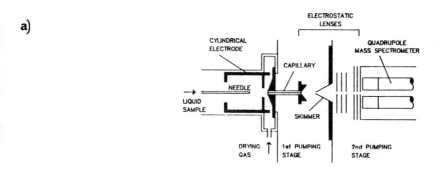

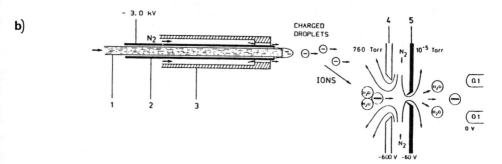

Figure 8.10. (a) Schematic of the electrospray interface. (Reprinted with permission from Ref. 36. Copyright 1985, American Chemical Society.) (b) Schematic of the ion spray interface: 1, 50 μm i.d. fused-silica capillary; 2, 0.20 mm id stainless steel capillary; 3, 0.8 mm i.d. Teflon tube with narrow bore insert; 4, ion focusing lens (counter electrode); 5, orifice holding plate with 100 μm i.d. conical orifice. (Reprinted with permission from Ref. 37. Copyright 1987, American Chemical Society.)

and because they both produce ions by desorption or emission of ions from small, highly charged evaporating droplets, they will be discussed together. As with thermospray MS, a dedicated ion source, in this case an API source, is required. In both techniques the sample effluent is passed through a small jet maintained at a potential of several kV. The voltage helps form and charge the droplets. A major difference between ESI and ISI is that in ISI there is an additional pneumatic nebulization stage to enhance droplet formation. Flow rates of 1–40 μL/min are typical for ESI, while ISI permits flow rates of up to 200 μL/min. High-aqueous-content mobile phases are compatible with ISI but not pure ESI. In ESI, however, a coaxial sheath flow containing the organic solvent can be used, which permits use of high-aqueous-content mobile phases. The use of these techniques is currently undergoing explosive growth, and many conflicting claims are being made. The utility of both is undisputed, however. In further discussion of these techniques, the term "ESI" will be used as a general acronym for both.

Two major advantages of the electrospray techniques have become apparent. First, these techniques are gentle and lead to less thermal degradation than is observed with some of the other techniques. A second advantage is that, often, the analyte ion contains multiple charges. This reduces the mass-to-charge ratio of the analyte and permits mass analysis of high-molecular-weight compounds on mass spectrometers with significantly lower mass ranges. For example, bovine albumin dimer (MW $\approx$ 133,000) yields a mass spectrum, with the largest molecular ion species having over 100 charges and appearing at an m/z value less than 1300.[38]

8.3. Applications

The purpose of this section is to give the reader an indication of what interfaces have been used, successfully and unsuccessfully (if reported), for different compound types, and to provide some indication of detection limits associated with the various interface types. This section will not serve as an exhaustive review of LC/MS applications. As with any article in a continually developing area that attempts to draw together numerous results from a variety of sources, there are a number of caveats to keep in mind. Because few laboratories have several different interfaces available, comparative studies are few. Negative results are less often published than are positive results. Rapidly developing areas often change so rapidly as to make a review or overview out of date as soon as it is written, and the writer is at the mercy of computerized literature searches, which have an annoying tendency of missing articles. Within these limitations, the author hopes that this section provides a fair overview of the scope of applications that have successfully used mass spectrometry as a detector for HPLC (considering the comparative costs and complexities of HPLC and MS, one could, and one usually does, argue as to the appropriateness of calling an MS an HPLC detector versus calling the HPLC a sample introduction system for the MS). For a more detailed discussion of applications, the reader is encouraged to look at current review articles and books (see for example, Refs. 3, 39–42). In addition a number of reviews have recently appeared that cover instrumental approaches to coupling LC with MS.[43–46]

8.3.1. Drugs and Drug Metabolites

The application of a variety of LC/MS techniques to the determination, both qualitative and quantitative, of drugs and their metabolites has received a significant amount of interest.[3,47–50] The types of drugs studied include steroids, alkaloids, and sulfa drugs, to mention a few. As many potential applications in this area are of proprietary compounds, this author assumes that a large number of applications have not been published in the open literature. Several of the drugs have been analyzed using a number of the interface types. A comparison of these results is useful for comparing the interfaces.

Ranitidine-N-oxide is a compound that is prone to thermal degradation and has been suggested as a test for thermal effects in LC/MS interfaces. This compound has been analyzed by the moving belt,[51] DLI,[6] thermospray,[52] and ESI[6] interfaces. The spectrum obtained from the moving belt interface exhibited the greatest amount of thermal degradation with no molecular ion and few high-mass ions observed. The DLI-produced spectrum contained a molecular ion of about 50% relative abundance with the base peak being due to the loss of oxygen. Both thermospray- (of recent design) and ESI-produced spectra with the molecular ion as the base peak. The TSP spectrum showed a greater oxygen loss than did the ESI spectrum. No particle-beam- or CF-FAB–produced spectra have been published.

Erythromycin has been analyzed using belt,[53] DLI,[6] TSP,[52] particle beam (onto a FAB probe),[24] ESI,[54] and CF-FAB[55] introduction of the sample. This compound, however, is thermally relatively stable. All interfaces gave spectra containing molecular ion species. Varying degrees of water loss were noted with ESI/ISI and TSP showing the lowest amount. Thus, all of these interfaces appear to work satisfactorily for this type of compound.

Due to the range of compound types included here, no generalization about the limits of detection can be made. Detection limits varying from the low picogram to low nanogram range were determined on the same instrument (ESI/ISI interface) for different compounds. This range of detection limits seems to be typical for a variety of drugs analyzed by LC/MS using different interfaces.

8.3.2. Herbicides, Pesticides, and Their Metabolites

Because of environmental concerns as well as the low thermal stability of many members of these classes, there has been a great deal of interest in applying LC/MS techniques to the analysis of herbicides, pesticides, and their metabolites.

A great deal of research has gone into the TSP/MS analyses of these compounds, possibly due to the more widespread availability of the technique. Certainly the results published indicate that TSP/MS is useful for these compounds. Molecular weight information can be obtained from the spectra of a wide variety of these compounds, including triazines,[56–58] carbamates,[59,60] organophosphorus compounds,[61–63] organochlorine compounds,[57,58,60] and sulfonylurea herbicides.[64] Detection levels in the low nanogram to low picogram region have been obtained for a number of these compounds.

Particle-beam LC/MS has begun to attract a lot of attention in this area also, now that commercial instruments are available. Budde and co-workers have investigated particle-beam MS (two different manufacturers) for the analysis of a number of carbamates, ureas, and thioureas.[22,65] A typical reconstructed total ion chromatogram (Figure 8.11) demonstrates that good chromatographic peak shape is retained in the particle-beam interface. The particle-beam MS was more likely to provide molecular weight information than was GC/MS analysis. One serious problem that was observed is the carrier effect. That is, enhancement of ion abundances were observed when compounds coeluted or when certain buffers such as ammonium acetate were used. Others have reported that quantification is affected by

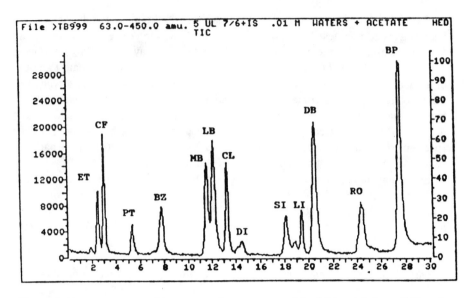

Figure 8.11. Reconstructed ion current chromatogram of the LC/particle-beam/MS analysis of ethylene thiourea (ET), caffeine (CF), *o*-chlorophenyl thiourea (PT), benzidine (BZ), 3,3'-dimethoxybenzidine (MB), 3,3'-dimethylbenzidine (LB), carbaryl (CL), diuron (DI), siduron (SI), linuron (LI), 3,3'-dichlorobenzidine (DB), rotenone (RO), and benzoylpropethyl (BP) using a 2 mm i.d. C18 column and 30:70 acetonitrile:0.01M aqueous ammonium acetate (1 min) programmed to 100:0 acetonitrile:0.01M aqueous ammonium acetate over 29 min. (Reprinted by permission of Elsevier Science Publishing Co., Inc. from Ref. 22, coypright 1990.)

the molecular weight of the compound and by changes in the solvent composition as observed in gradient runs.[66] Thus, more work is needed to verify the utility of particle-beam MS for quantitative analysis and for low-level qualitative analysis.

Using DLI and nanoscale capillary columns, picogram and femtogram sensitivities were observed for a number of trifluralin metabolites,[67] as well as for other pesticides and herbicides including metribuzin[68] and urea- and carbamate-based compounds.[69] A potential problem with these compounds is that, if a significant amount of matrix material is present, the nature of the chromatographic stationary phase may change.

Neither CF/FAB nor ESI has been used extensively for the analysis of this group of compounds. In those cases that have been reported, reasonable data have been presented.[70,71] Not surprisingly, CF/FAB has had the best results when looking at polar conjugates.

Thus, while a number of LC/MS interfaces have been successfully used for the analysis of herbicides, pesticides, and their metabolites, TSP/MS has certainly had the greatest success for this class of compounds.

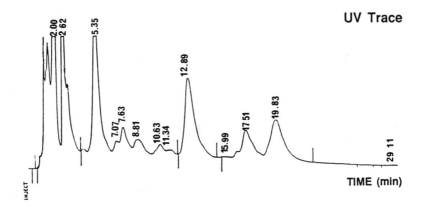

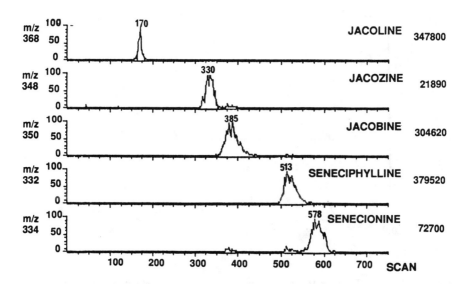

Figure 8.12. On-line HPLC UV trace and reconstructed ion chromatograms of alkaloids from a *Senecio jacobaea* extract using a gradient of 10 to 30% acetonitrile in 0.1*N* ammonium hydroxide over 20 min. (Reprinted from Ref. 81, John Wiley & Sons, Ltd., copyright 1990.)

8.3.3. Natural Products

This category of compounds is rather a catchall for a variety of structurally unrelated compounds including tricothecenes,[72,73] alkaloids,[54,74–76] and glucosinolates.[77–80] TSP/MS has been most often applied to these problems with relatively good success.

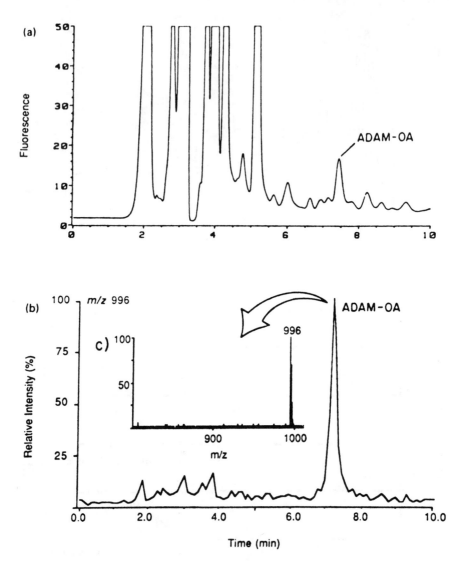

Figure 8.13. (a) Isocratic HPLC analysis (10:90 aqueous acetonitrile on a 2.1 mm i.d. RP-18 column) of Bay of Gaspe plankton extract derivatized with anthryldiazomethane (ADAM), (b) extracted ion chromatogram of *m/z* 966 of the ADAM derivative of okadaic acid (OA) from an ion spray source, and (c) mass spectrum of ADAM-OA. (Reprinted with permission from Ref. 84. Copyright 1990, John Wiley & Sons, Ltd.)

Detection limits for tricothecenes were in the low ng range for full scan data, in the 50–500 pg range by selected ion monitoring (SIM)[72]. Also, detection levels of under 10 pg by SIM were reported for selected alkaloids.[54]

Applications of particle beam, CF-FAB, or ESI have been less often reported. It has been noted that EI spectra of pyrrolizidine alkaloids with low ng detection limits have been obtained by the particle-beam method.[80] LC TSP/MS separation and identification of the pyrrolizidine alkaloids and their N-oxides isolated from plants have also been reported.[81] A comparison of the on-line HPLC/UV chromatogram and the selected ion chromatograms of the alkaloid components of *Senecio Jacobea* (Figure 8.12) illustrates the excellent chromatographic fidelity obtained with TSP/MS using a 4.3 mm i.d. column and gradient conditions. Continuous flow and Frit-FAB have been employed for the determination of glucosinolates with LODs (at the MS) of less than 1 ng[78] and of saponins.[82] Quilliam and co-workers have reported the determination of shellfish toxins by ESI with low pg detection levels, which are sufficient for detection of the toxin in biological samples (Figure 8.13).[83,84]

8.3.4. Environmental Samples

The most thoroughly studied group of compounds in this category is dyes. The analysis of dyes by LC/MS methodologies has been of considerable interest due to the difficulty of obtaining spectral data from these compounds by more conventional inlet systems and because of their importance in environmental analyses.

TSP/MS and ESI/MS have been most successful for dye analysis. Voyksner used LC TSP/MS to separate and obtain spectra of azo and diazo dyes in gasoline, and disperse dyes spiked into waste water and soil samples.[85] Betowski and Ballard were also able to obtain spectra of a number of dye classes, including monosulfonated azo dyes, by TSP/MS.[86,87] Using optimized TSP parameters under negative ion conditions, Flory and co-workers obtained molecular ion species for di-, tri-, and tetra-sulfonated dyes.[88] With further improvements to the TSP source, including the use of a wire repeller, limits of detections were decreased to 0.05–20 ng.[89] In the latter study, positive ion TSP spectra were obtained for the first time. Freas has reported the optimization of an LC/TSP/MS system for sulfonated dyes with results similar to that of Yinon et al.[90]

Bruins et al. obtained negative ion ESI spectra of sulfonated azo dyes separated by HPLC.[37] Molecular ion species were the most abundant ions observed for most examples. When introduced to CZE rather than by HPLC, multiply charged species were obtained as the most abundant ions.[91]

Yinon and co-workers also investigated the utility of particle-beam MS for the analysis of azo-, arylmethyl-, anthraquinone-, coumarin-, and xanthone-based dyes.[92] EI-type spectra were obtained, and molecular ions were usually observed, but detection levels were 2–3 orders of magnitude worse than observed with their TSP/MS system.[89]

Aromatic sulfonic acids are a class of compounds closely related to dyes. Brown and co-workers have used LC–particle-beam–MS successfully to characterize the aromatic sulfonic acids in leachates from a hazardous waste disposal site.[93,94]

Budde and co-workers have investigated the utility of TSP and particle beam for the determination of diphenylamine and N-nitrosodiphenylamine. LODs by TSP by 50–150 ng, while N-nitrosodiphenylamine could not be detected at the 500 ng level.[22,95]

8.3.5. Peptides and Proteins

The LC/MS determination of high-molecular-weight peptides and small proteins is an area that is currently undergoing a tremendous explosion in interest and in development. Both the moving belt and DLI have been used for peptide analysis. The moving belt could only be used with derivatized peptides, however. An early problem noticed with DLI was variable tuning, where the mass spectrum can change significantly depending upon instrumental conditions. In this context, tuning refers to interface conditions, not source conditions. In an investigation of a series of peptides, $(Gly)_n$ with $n = 2$–6, under constant conditions, $(M + H)^+$ ions were observed for $n = 2$–4 but not for $n = 5$ and 6.[96] Peptides of this size are amenable to DLI analysis, as evidenced by the successful determination of Met- and Leu-enkephalin.[97]

Frit FAB[27] and CF-FAB[29] were the first LC/MS interfaces that were extensively used for peptide analyses. The CF-FAB interfaces were observed to provide improved sensitivity due to lowered matrix concentration and to less pronounced hydrophilicity/hydrophobicity effects. Caprioli, for example, has reported a 200-fold increase in sensitivity for peptides in the 1500 dalton mass range at the picomole/femtomole level.[98] Using nanoscale columns, attomole sensitivity has been achieved for selected compounds.[32] Extensive use of these techniques is being made in the FAB mapping of tryptic digests (see, for example, Refs. 99–107). For example, using this technique, subpicomole amounts of known peptides have been analyzed,[108] while over 90% of the expected tryptic fragments from 30 picomoles of lys-plasminogen were detected using frit-FAB.[109]

The particle-beam interface has been integrated with FAB for the analysis of peptides that could not be analyzed by typical particle-beam conditions.[24] In this mode the particle beam impinges upon a static FAB probe. The analyte is then desorbed in the traditional manner by FAB. The initial experiments with this combination used flow injection analysis. Advantages reported for the combination include higher flow rates and the use of nonaqueous LC solvents such as hexane. To be practical for use with extended HPLC analyses, however, the FAB matrix will need to be renewed periodically. This requirement will probably be best met by incorporating a continuous-flow FAB probe.

TSP/MS has also been utilized for the determination of peptides, but not as extensively as have the FAB techniques. Variability of tuning conditions has been observed to be a problem with the TSP interface similar to that observed for DLI.[110] TSP/MS has also been used in conjunction with immobilized enzymes for the determination of peptide sequences.[111,112] Twenty-nine of 58 amino acid residues in the protein basic pancreatic trypsin inhibitor have been identified in this manner. Voyksner has reported sensitivities of less than 1 pm in a study of the on-line immobilized enzyme hydrolysis of neuropeptides by TSP.[113] Blackstock et al. have confirmed the complete sequence of recombinant human interleukin-2 expressed by *Escherichia coli* by LC TSP/MS of the tryptic digest.[114] Straub and Chan have reported the use of TSP to generate multiply charged ions from proteins. The sensitivity was reported to be lower than with ESI, however.[115,116] Pramanik and co-workers have used LC/TSP/MS for the determination of the amino acid content of peptides by forming either the phenylthiocarbamyl or phenylthiohydantoin derivatives.[117,118]

The most exciting recent LC/MS application to peptides and proteins has involved ESI/MS. Although multiply charged ions had been observed in ESI spectra earlier, the application of this phenomenon to proteins has occurred only recently. Since the first observations by Fenn and co-workers, Smith and co-workers, and Henion and co-workers in 1988 and 1989,[38,119,120] the ESI/MS spectra of a number of proteins with molecular weights up to 133 kilodaltons have been reported. Sciex has reported the use of HPLC (1 mm columns)/MS for the separation and identification of components of tryptic digests and of mixtures of peptides and proteins with molecular weights up to 20 kDa.[35] A number of groups have now reported the separation and analysis of peptide mixtures and tryptic digests by LC/ESI MS using 1 mm columns,[106,121,122] 250 μm columns,[123] and 75 μm columns.[107,108] Henion has reported detection limits of 65 attomole for leucine enkephalin by ESI with a linear response from 29 fm to 90 pm on column (Figure 8.14).[121] As the availability of ESI/ISI instrumentation increases, the number and range of reported applications is expected to increase significantly.

The report by Hemling et al. contains an evaluation of CF-FAB and ESI for sensitivity, coverage, discrimination effects, and utility for glycopeptide determina-

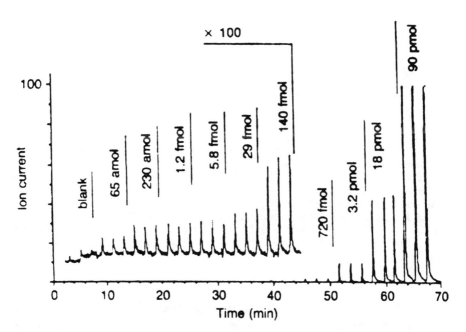

Figure 8.14. Total ion chromatogram of triplicate on-column injections of leucine enkephalin from 65 amol to 90 pmol using an ion spray source. (Reprinted by permission of MicroSeparations, Inc. from the *Journal of Microcolumn Separations*, Ref. 121.)

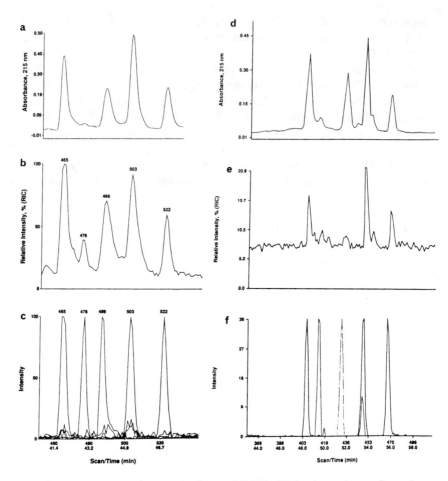

Figure 8.15. Separation of a tryptic digest of RCM sCD4 using a 1 mm C_{18} column. Mobile phases for the CF-FAB analyses are: (A) water/glycerol 95:5 with 0.1% tri-fluoroacetic acid (TFA), (B) acetonitrile/water/glycerol 60:35:5 containing 0.1% TFA; and for the ESI experiments; Mobile phase; 0.1% TFA (Mobile phase A) and acetonitrile/water 90:10 containing 0.1% TFA (Mobile phase B) (a) UV trace, (b) CF-FAB RIC, and (c) CF-FAB selected ion chromatograms using 7% B (5 min) to 19% B over 11 min; (d) UV trace, (e) ESI RIC, and (f) ESI selected ion chromatograms using 5% B (5 min) to 18% B (18 min) to 45% B (54 min) and then to 90% B (32 min). (Reprinted from Biomedical and Environmental Mass Spectrometry (Ref. 106, ©John Wiley & Sons, Ltd., 1990) by permission of John Wiley & Sons, Ltd.)

tion.[106] A comparison of the UV trace and reconstructed ion chromatograms from CF-FAB and ESI analyses is shown in Figure 8.15. A better chromatographic separation was obtained with LC/ESI/MS, which the authors attributed to the necessity of adding glycerol to the mobile phase. Application of the coaxial approach[30-33] or postcolumn addition[101] for delivery of the FAB matrix should eliminate this problem, however. The authors concluded, based on the less limited mass range of ESI due to multiple charging and lack of high mass discrimination, that ESI also delivers better sensitivity.

8.3.6. Nucleobases, Nucleosides, and Nucleotides

There have been reports published on the application of all of the LC/MS techniques except particle beam to the analysis of these compounds.[36,40,54,55,124–130] The spectra obtained from nucleobases tend to be dominated by molecular ion species, while varying ratios of protonated molecular ion and fragment ions due to the protonated base or sugar moiety are observed in the spectra of nucleosides. The ratio of parent to fragment ions in the spectra is often dependent on source and interface conditions, especially for DLI and TSP/MS. Reported sensitivities for ESI and TSP are in the 0.1 to 5 ng range for nucleosides under SIM conditions.[54,130] Nucleotides yield more complex spectra and often relatively low-abundance or nonobservable molecular ion species.

LC TSP/MS has been successfully applied to the analysis of nucleosides and bases in DNA and RNA digests[126,127,130,131] (Figure 8.16). Sensitivities of 20 ng per nucleoside in 2.5 to 30 μg of crude digest have been reported. Although this is sufficient to detect some naturally occurring methylated bases, it is insufficient to analyze many modified bases at physiologically important levels. With suitable derivatization, however, the sensitivities of moving belt LC/MS for 5-methylcytosine and 5-hydroxymethyluracil (9.9 pg and 180 fg, respectively) approached that of GC/MS.[132]

The recent increased interest in the application of ESI/MS techniques to high-mass substances has also been apparent in the area of oligonucleotide analysis. For example, Henion has reported the spectrum of a 14-mer (MW = 4260) by ESI, in which the base peak is the $(M - 9H)^-$ anion.[120] At this time, however, little has been done with preseparation by LC.

Millington and co-workers have explored the utility of CF-FAB for the determination of acyl carnitines and acyl coenzyme-A compounds, which are markers for a number of metabolic diseases.[133,134] Detection limits suitable for metabolic profiling (1 nm/g for the carnitines and 50–100 pm for the Co-A compounds) were observed.

8.3.7. Saccharides

This class of compounds has proved to be the most refractive to analysis by LC/MS. Typically, most reports have been on the analysis of trisaccharides and smaller.[124,135,136] Under TSP/MS conditions, molecular ions of underivatized species are of relatively low abundance. O-Methylation significantly increases sensitivity and molecular ion abundance for TSP/MS analysis.[137] Recently, a method for coupling of alkaline anion exchange LC with TSP/MS, which incorporates a membrane suppressor to remove nonvolatile alkaline salts, has been reported.[138] Detection limits in the μg range were reported for N-acetylated mono- and disaccharides by this technique.

CF-FAB techniques have been applied with more success.[32,139] Underivatized

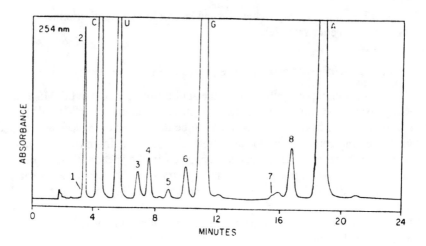

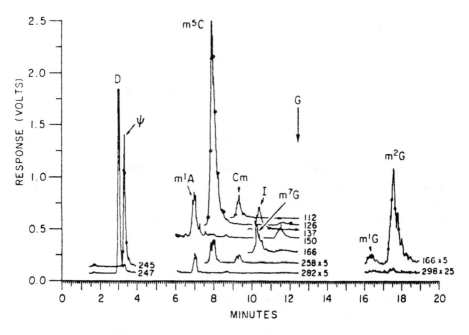

Figure 8.16.   Separation of nucleosides from 2.5 μg of rabbit liver tRNA^{Val} separated on a 4.0 mm μBondapack column and a linear solvent program from 5% to 15% methanol over 30 min. (a) UV trace, and (b) selected ion chromatograms of $(M + H)^+$ and $(BH_2)^+$ ions of nucleosides. (Reprinted from Ref. 130 by permission of Oxford University Press.)

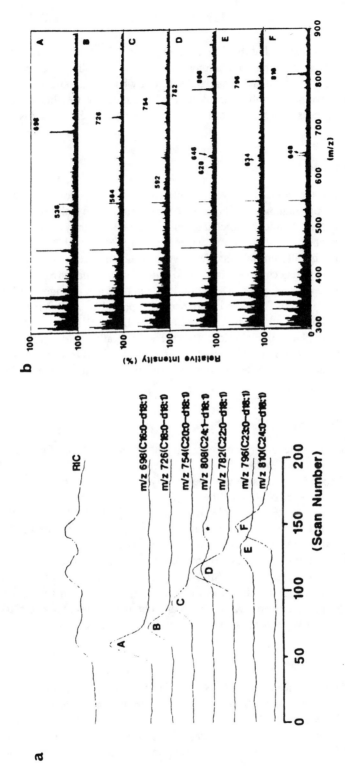

Figure 8.17. (a) Mass chromatograms of the molecular ion species of glucosylceramide purified from the spleen of a patient with Gaucher's disease using LC/Frit-FAB/MS. (b) Mass spectra of the molecular ion species observed in (a). (Reprinted with permission from Ref. 147.)

187

maltoheptose was analyzed at the 500 ng level with CF-FAB, and, using nanoscale capillary introduction, 5 ng of maltotetrose was successfully analyzed.[32] With permethylation, increased mass range and sensitivity have been observed.[139] Thus, permethylated ovomucoid oligosaccharides with molecular weights up to 3086 daltons were successfully separated and analyzed using a packed fused-silica column.

8.3.8. Lipids

Major interest in the lipid category has centered on the separation and analysis of eicosanoids and prostaglandins. The reported applications have almost exclusively been with TSP/MS. Although molecular ion species are normally present, the base peak in the spectrum is often due to water loss(es).[100,135,140–146] Differences in spectra obtained in different laboratories have been noted. These differences could be due either to differences in "tuning" or to improvements that have been made to the TSP interface over the years. Detection levels in the low ng range by SIM have been reported.[135,140,142] This is in comparison to reported GC/MS detection levels of mid femtograms. Suzuki et al. have used Frit-FAB with 1 mm columns for the determination of glycosphingolipids.[147,148] The separation and analysis of the glucosylceramides identified from 1 μg of ceramides purified from the spleen of a patient with Gaucher's disease illustrate the utility of the technique for clinical applications (Figure 8.17).

8.3.9. Phospholipids

Primary emphasis in the majority of reports on the LC/MS analysis of phospholipids has been on qualitative rather than quantitative analyses.[17,140,149,150] TSP/MS analysis has been used extensively in these analyses. In a comparison of data from different laboratories, it is obvious that interface tuning is a crucial component in obtaining good spectra. For example, Kim and co-workers[140,150] obtained a 10% relative abundance of $(M + H)^+$ ion from 16:0,18:2-PC, while Mallet and Rollins[149] obtained the $(M + H)^+$ ion as the base peak in the spectrum of 14:0,14:0-PC by tuning specifically on the protonated molecular ion. Kim, Yergey, and Salem have also obtained LC TSP/MS separation and analysis of phosphatidyl ethanolamines as well as cholines from bovine heart.[146] Sherman and co-workers studied inositol phosphates by LC/TSP/MS and obtained LODs of 100 pm/μL that were too high for a general method for biological applications.[151]

8.3.10. Steroids

A number of groups have employed TSP/MS for the analysis of steroids and steroid conjugates. SIM detection levels of 100 pg are generally achievable.[152–156] As with the phospholipids discussed previously, tuning can be critical for the successful analysis of low levels of steroids.[153] Gaskell and co-workers have rigorously compared LC TSP/MS and GC/MS for the quantitative analysis of serum cholesterol

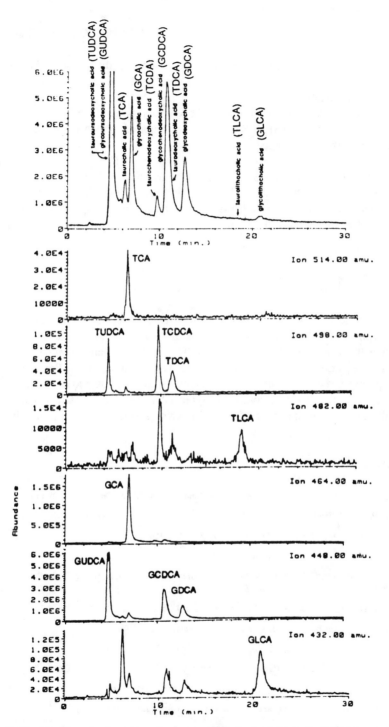

Figure 8.18. Reconstructed total ion chromatogram and selected ion chromatograms of the biliary bile acids from a patient with gallstones after oral bile acid therapy with ursodeoxycholic acid analyzed on 460 mm Ultrasphere ODS column using an isochratic (methanol–0.4*M* ammonium acetate at 1.0 mL/min) elution and TSP/MS. (Reprinted with permission from Ref. 161.)

levels.[157] TSP/MS detection levels were 10-fold higher than those for GC/MS, and quantitation was less precise as well. Similar detection levels and precision was observed by Estaban et al. [158]

SIM detection levels closer to 10 pg have been reported for ESI/MS.[54,159] Better precision compared to TSP/MS has also been reported, possibly due the tendency of TSP/MS peaks to be jagged.

Steroid conjugates have successfully been analyzed by both TSP/MS (glucuronides)[155] and by ESI/MS (glucuronides and sulfates),[159] while bile salts have been successfully analyzed by CF-FAB/MS.[160] Setchell and Vestal have demonstrated the application of TSP/MS for the determination of bile acids in patients with gall stones (Figure 8.18) and other diseases that alter the natural bile acid profile.[161]

8.4. Conclusion

The purpose of this chapter has been to give an overview of the many modern variations of LC/MS interfaces available and an indication of the range of compounds that can be analyzed by LC/MS. The field is still undergoing rapid change. At the present time no one interface is optimum for all analyses. Possibly, in time, one interface will dominate, much as capillary GC has come to dominate GC/MS. Given the wider range of compound types addressed by LC/MS, it would not be totally unexpected, however, if several interfaces obtain long-term viability.

8.5. References

1. R. P. W. Scott, C. J. Scott, M. Munroe, and J. Hess, Jr., *J. Chromatogr.* **99** (1974) 395.

2. W. H. McFadden, H. L. Schwartz, and A. Evans, *J. Chromatogr.* **122** (1976) 389.

3. D. E. Games, *Biomed. Mass Spectrom.* **8** (1981) 454.

4. P. Arpino, *Mass Spectrom. Rev.* **8** (1989) 35.

5. J. G. Stroh, J. C. Cook, R. M. Milberg, L. Brayton, T. Kiharo, Z. Huang, K. L. Rinehart, Jr., and I. A. S. Lewis, *Anal. Chem.* **57** (1985) 985.

6. J. D. Henion, T. R. Covey, D. Silvestre, and J. B. Crowther, *LC* **3** (1985) 240.

7. J. W. Jorgenson and E. J. Guthrie, *J. Chromatogr.* **255** (1983) 335.

8. W. M. A. Niessen and H. Poppe, *J. Chromatogr.* **385** (1987) 1.

9. J. S. M. de Wit, C. E. Parker, K. B. Tomer, and J. W. Jorgenson, *Anal. Chem.* **59** (1987) 2400.

10. H. Alborn and G. Stenhagen, *J. Chromatogr.* **323** (1985) 47.

11. P. Hirter, H. J. Walther, and P. Daetwyler, *J. Chromatogr.* **323** (1985) 89.

12. G. Stenhagen and H. Alborn, *J. Chromatogr.* **474** (1989) 285.

13. C. R. Blakley and M. L. Vestal, *Anal. Chem.* **55** (1985) 750.

14. D. Zakett, G. J. Kallos, and P. J. Savicksa, 32nd Annual Conference on Mass Spectrometry and Allied Topics, May 27–June 1, 1984, San Antonio, TX, p. 3.

15. W. H. McFadden and S. A. Lammert, *J. Chromatogr.* **385** (1987) 201.

16. W. M. A. Niessen, R. A. M. van der Hoeven, M. A. G. de Kraa, C. E. M. Heeremans, U. R. Tjaden, and J. van der Greef, *J. Chromatogr.* **478** (1989) 325.

17. H. Y. Kim and N. Salem, Jr., *Anal. Chem.* **58** (1986) 9.

18. R. C. Willoughby and R. F. Browner, *Anal. Chem.* **56** (1984) 2625.

19. Extrel Corporation, Pittsburgh, PA.

20. R. F. Browner, 6th (Montreux) Symposium on Liquid Chromatography/Mass Spectrometry, Ithaca, NY, July 19–21, 1989.

21. R. D. Voyksner, C. S. Smith, and P.C. Knox, *Biomed. Environ. Mass Spectrom.* **19** (1990) 523.

22. T.A. Bellar, T. D. Behymer, and W. L. Budde, *J. Am. Soc. Mass Spectrom.* **1** (1990) 92.

23. Discussion at 7th (Montreux) Symposium on Liquid Chromatography/Mass Spectrometry, Montreux, Switzerland Oct. 31–Nov. 2, 1990.

24. J. D. Kirk and R. F. Browner, *Biomed. Environ. Mass Spectrom.* **18** (1989) 355.

25. Vestec Corporation, Houston, TX.

26. M. Barber, R. S. Bordoli, R. D. Sedwick, and A. N. Tyler, *J. Chem. Soc., Chem. Commun.,* **1981,** 325.

27. Y. Ito, T. Takeuchi, D. Ishi, and M. Goto, *J. Chromatogr.* **346** (1985) 161.

28. R. M. Caprioli, T. Fan, and J. S. Cottrell, *Anal. Chem.* **58** (1986) 2949.

29. R. M. Caprioli, *Anal. Chem.* **62** (1990) 477a.

30. R. M. Caprioli, ed., *Continuous-Flow Fast Atom Bombardment Mass Spectrometry,* John Wiley & Sons: Chichester (1990).

31. J. S. M. de Wit, L. J. Deterding, M. A. Moseley, K. B. Tomer, and J. W. Jorgenson, *Rap. Commun. Mass Spectrom.* **2** (1988) 100.

32. M. A. Moseley, L. J. Deterding, J. S. M. de Wit, K. B. Tomer, R. T. Kennedy, N. Bragg, and J. W. Jorgenson, *Anal. Chem.* **61** (1989) 1577.

33. S. Pleasance, P. Thibault, M. A. Moseley, L. J. Deterding, K. B. Tomer, and J. W. Jorgenson, *J. Am. Soc. Mass Spectrom.* **1** (1990) 312.

34. J. B. Fenn, M. Mann, C. K. Meng, and S. F. Wong, *Mass Spectrom. Rev.* **9** (1990) 57.

35. Sciex, The API III Book, Sciex Corporation, Thornhill, Ontario, Canada (1989).

36. C. M. Whitehouse, R. N. Dreyer, M. Yamashita, and J. B. Fenn, *Anal. Chem.* **57** (1985) 675.

37. A. P. Bruins, L. O. G. Weidolf, J. D. Henion, and W. L. Budde, *Anal. Chem.* **59** (1987) 2642.

38. J. A. Loo, H. R. Udseth, and R. D. Smith, *Anal. Biochem.* **179** (1989) 404.

39. K. B. Tomer and C. E. Parker, *J. Chromatogr.* **492** (1989) 189.

40. A. L. Yergey, C. G. Edmonds, I. A. S. Lewis, and M. L. Vestal, *Liquid Chromatography/Mass Spectrometry Techniques and Applications,* Plenum: New York (1990).

41. *Liquid Chromatography/Mass Spectrometry,* ACS Symposium Ser., **420,** American Chemical Society: Washington, DC (1990).

42. J. Henion and E. Lee, *Pract. Spectrosc.* **8** (1990) 469–503.

43. E. D. Lee and J. D. Henion, *J. Chromatogr. Sci.* **23** (1985) 253.

44. C. G. Edmonds, J. A. McCloskey, and V. A. Edmonds, *Biomed. Environ. Mass Spectrom.* **10** (1983) 237.

45. T. Takeuchi, *Fres. J. Anal. Chem.* **337** (1990) 631.

46. P. Arpino, *Fres. J. Anal. Chem.* **337** (1990) 667.

47. K. M. Straub, in *Progress in Drug Metabolism* **11**, G. G. Gibson, ed., 267, Taylor & Francis: London (1988).

48. K. Vekey, D. Edwards, and L. F. Zerilli, *J. Chromatogr.* **488** (1989) 73.

49. L. D. Bowers, *Clin. Chem.* **35** (1989) 1282.

50. P. Kokkonen, E. Schroeder, W. M. A. Niessen, U. R. Tjaden, and J. van der Greef, *J. Chromatogr.* **551** (1990) 35.

51. L. E. Martin, J. Oxford, and F. J. N. Tanner, *Xenobiotica* **11** (1981) 831f.

52. Kratos Analytical, MS25 LC/MS Data Compilation, Ramsey, NJ, 1985.

53. P. J. Arpino and G. Guiochon, *Anal. Chem.* **51** (1979) 683A.

54. M. Sakairi and H. Kambara, *Anal. Chem.* **60** (1988) 774.

55. Kratos Analytical, Continuous Flow Fast Atom Bombardment, Ramsey, NJ, 50 pp.

56. D. Barcelo, *Org. Mass Spectrom.* **24** (1989) 219.

57. D. Barcelo, *Org. Mass Spectrom.* **24** (1989) 898.

58. R. D. Voyksner and C. A. Haney, *Anal. Chem.* **57** (1985) 991.

59. K. S. Chiu, A. van Langenhove, and C. Tanaka, *Biomed. Environ. Mass Spectrom.* **18** (1989) 200.

60. I. Hammond, K. Moore, H. James, and C. Watts, *J. Chromatogr.* **474** (1989) 175.

61. D. Barcelo, *Biomed. Environ. Mass Spectrom.* **17** (1989) 363.

62. E. R. J. Wils and A. G. Hulst, *J. Chromatogr.* **454** (1988) 261.

63. D. Barcelo and J. Albaiges, *J. Chromatogr.* **474** (1989) 163.

64. A. C. Barefoot and R. W. Reiser, *Biomed. Environ. Mass Spectrom.* **18** (1989) 77.

65. T. D. Behymer, T. A. Bellar, and W. L. Budde, *Anal. Chem.* **62** (1990) 1686.

66. 6th (Montreux) Symposium on Liquid Chromatography/Mass Spectrometry, Ithaca, NY, July 19–21, 1989.

67. J. S. M. de Wit, C. E. Parker, J. W. Jorgenson, and K. B. Tomer, *Biomed. Environ. Mass Spectrom.* **17** (1988) 47.

68. B. H. Escoffier, C. E. Parker, T. C. Meter, J. S. M. de Wit, F. T. Corbin, J. W. Jorgenson, and K. B. Tomer, *J. Chromatogr.* **474** (1989) 301.

69. J. R. Perkins, C. E. Parker, and K. B. Tomer, *Proceedings of the 38th ASMS Conference on Mass Spectrometry and Allied Topics,* Tucson, AZ, June 3–8, 1079 (1990).

70. K. W. M. Siu, G. J. Gardner, and S. S. Berman, *Rap. Commun. Mass Spectrom.* **2** (1988) 69.

71. A. C. Barefoot, R. W. Reiser, and S. A. Cousins, *J. Chromatogr.* **474** (1989) 39.

72. R. D. Voyksner, W. M. Hagler, Jr., K. Tyczkowska, and C. A. Haney, *J. High Res. Chrom. Commun.* **8** (1985) 119.

73. T. Krishnamurthy, D. J. Beck, R. K. Isensee, and B. B. Jarvis, *J. Chromatogr.* **469** (1989) 209.

74. S. Auriola, T. Naaranlahti, R. Kostiainen, and S. P. Lapinjoki, *Biomed. Environ. Mass Spectrom.* **19** (1989) 400.

75. S. Auriola, T. Naaranlahti, R. Kostiainen, and S. P. Lapinjoki, *Biomed. Environ. Mass Spectrom.* **19** (1989) 609.

76. S. Auriola, T. Naaranlahti, R. Kostiainen, and S. P. Lapinjoki, *J. Chromatogr.* **474** (1989) 181.

77. L. R. Hogge, D. W. Reed, E. W. Underhill, and G. W. Haughn, *J. Chromatogr. Sci.* **26** (1988) 551.

78. P. Kokkonen, J. van der Greef, W. M. A. Niessen, U. R. Tjaden, G. J. Ten Hove, and G. van de Werken, *Rap. Commun. Mass Spectrom.* **3** (1989) 102.

79. C. E. M. Heeremans, R. A. M. van der Hoeven, W. M. A. Niessen, J. Vuik, R. H. de Vos, and J. van der Greef, *J. Chromatogr.* **472** (1989) 219.

80. W. E. Harris and R. F. Browner, *Proceedings of the 36th ASMS Conference on Mass Spectrometry and Allied Topics,* 1069 (1988).

81. C. E. Parker, S. Verma, K. B. Tomer, R. L. Reed, and D. R. Buhler, *Biomed. Environ. Mass Spectrom.* **19** (1990) 1.

82. M. Hattori, Y. Kawata, N. Kakiuchi, K. Matsuura, T. Tomimori, and T. Namba, *Chem. Pharm. Bull.* **36** (1988) 4467.

83. M.A. Quilliam, B. A. Thompson, G. J. Scott, and K. W. M. Siu, *Rap. Commun. Mass Spectrom.* **3** (1989) 145.

84. S. Pleasance, M. A. Quilliam, A. S. W. de Freitas, J. C. Marr, and A. D. Cembella, *Rap. Commun. Mass Spectrom.* **4** (1990) 206.

85. R. D. Voyksner, *Anal. Chem.* **57** (1985) 2600.

86. L. D. Betowski and J. M. Ballard, *Anal. Chem.* **56** (1984) 2604.

87. J. M. Ballard and L. D. Betowski, *Org. Mass Spectrom.* **21** (1986) 575.

88. D. A. Flory, M. M. McLean, M. L. Vestal, and L. D. Betowski, *Rap. Commun. Mass Spectrom.* **1** (1987) 48.

89. J. Yinon, T. L. Jones, and L. D. Betowski, *Biomed. Environ. Mass Spectrom.* **18** (1989) 445.

90. M. A. McLean and R. B. Freas, *Anal. Chem.* **61** (1989) 2054.

91. E. D. Lee, W. Mueck, J. D. Henion, and T. R. Covey, *Biomed. Environ. Mass Spectrom.* **18** (1989) 233.

92. J. Yinon, T. L. Jones, and L.D. Betowski, *J. Chromatogr.* **482** (1989) 75.

93. M. A. Brown, I. S. Kim, R. Roehl, F. I. Sasinos, and R. D. Stephens, *Chemosphere* **19** (1990) 1921.

94. I. S. Kim, F. I. Sasinos, R. D. Stephens, and M. A. Brown, *Environ. Sci. Technol.* **24** (1990) 1832.

95. J. S. Ho, T. A. Bellar, J. W. Eichelberger, and W. L. Budde, *Environ. Sci. Technol.* **24** (1990) 1748.

96. H. Milon and H. Bur, *J. Chromatogr.* **350** (1985) 399.

97. C. N. Kenyon, *Biomed. Environ. Mass Spectrom.* **10** (1983) 535.

98. R. M. Caprioli and T. Fan, *Biochem. Biophys. Res. Commun.* **141** (1986) 1058.

99. R. M. Caprioli, in *Methods in Enzymology,* J. D. McCloskey, ed., Academic: New York (1990).

100. R. M. Caprioli, B. B. DaGue, and K. Wilson, *J. Chromatogr. Sci.* **26** (1988) 640.

101. J. E. Coutant, T.-M. Chen, and B. L. Ackermann, *J. Chromatogr.* **529** (1990) 265.

102. R. M. Caprioli, W. T. Moore, B. DaGue, and M. Martin, *J. Chromatogr.* **443** (1988) 355.

103. K. Mock, J. Firth, and J. S. Cottrell, *Org. Mass Spectrom.* **24** (1989) 591.

104. J. T. Stults, J. H. Bourell, E. Canova-Davis, V. T. Ling, G. R. Laramee, J. W. Winslow, P. R. Griffin, E. Rinderknecht, and R. L. Vandlen, *Biomed. Environ. Mass Spectrom.* **19** (1990) 65.

105. W. J. Henzel, J. H. Bourell and J. T. Stults, *Anal. Biochem.* **187** (1990) 228.

106. M. E. Hemling, G. D. Roberts, W. Johnson, S. A. Carr, and T. R. Covey, *Biomed. Environ. Mass Spectrom.* **19** (1990) 677.

107. L. J. Deterding, C. E. Parker, J. R. Perkins, M. A. Moseley, J. W. Jorgenson, and K. B. Tomer, *J. Chromatogr.*, **554** (1991) 329.

108. M. A. Moseley, L. J. Deterding, K. B. Tomer, and J. W. Jorgenson, *Anal. Chem.* **63** (1991) 1467.

109. D. J. Bell, M. D. Brightwell, W. A. Neville, and A. West, *Rap. Commun. Mass Spectrom.* **4** (1990) 88.

110. D. E. Games and E. D. Ramsey, *J. Chromatogr.* **323** (1985) 67.

111. H. Y. Kim, D. Pilosof, D. K. Dyckes, and M. L. Vestal, *J. Am. Chem. Soc.* **106** (1984) 7304.

112. K. Stachowiak, C. Wilder, M. L. Vestal, and D. F. Dyckes, *J. Am. Chem. Soc.* **110** (1988) 1758.

113. R. D. Voyksner, D. C. Chen, and H. E. Swaisgood, *Anal. Biochem.* **188** (1990) 72.

114. W. P. Blackstock, R. J. Dennis, S. J. Lane, J. I. Sparks, and M. P. Weir, *Anal. Biochem.* **175** (1988) 319.

115. K. Chan, D. Wintergrass, and K. Straub, *Rap. Commun. Mass Spectrom.* **4** (1990) 139.

116. K. Straub and K. Chan, *Rap. Commun. Mass Spectrom.* **4** (1990) 267.

117. B. C. Pramanik, C. R. Moomaw, C. T. Evans, S. A. Cohen, and C. A. Slaughter, *Anal. Biochem.* **176** (1989) 269.

118. B. C. Pramanik, S. M. Hinton, D. S. Millington, T. A. Dourdeville, and C. A. Slaughter, *Anal. Biochem.* **175** (1989) 305.

119. C. K. Meng, M. Mann, and J. B. Fenn, *Zeit. Phys. D* **10** (1988) 361.

120. T. R. Covey, R. F. Bonner, B. I. Shushan, and J. D. Henion, *Rap. Commun. Mass Spectrom.* **2** (1988) 249.

121. E. D. Lee, J. D. Henion, and T. R. Covey, *J. Microcol. Sep.* **1** (1989) 14.

122. E. C. Huang and J. D. Henion, *J. Am. Soc. Mass Spectrom.* **1** (1990) 158.

123. M. Hail, S. Lewis, I. Jardine, J. Liu, and M. Novotny, *J. Microcol. Sep.* **2** (1990) 285.

124. D. E. Games, M. A. McDowall, K. Levsen, K. H. Schager, P. Dobberstein, and J. L. Gower, *Biomed. Environ. Mass Spectrom.* **11** (1984) 87.

125. E. L. Esmans, Y. Luyten, and F. C. Alderweireldt, *Biomed. Environ. Mass Spectrom.* **10** (1983) 347.

126. E. L. Esmans, P. Geboes, Y. Luyten, and F. C. Alderweireldt, *Biomed. Environ. Mass Spectrom.* **12** (1985) 241.

127. F. C. Alderweireldt, E. L. Esmans, and P. Geboes, *Nucleosides Nucleotides* **4** (1985) 135.

128. K. J. Volk, M. S. Lee, R. A. Yost, and A. Brajter-Toth, *Anal. Chem.* **60** (1988) 722.

129. P. A. Blau, J. W. Hines, and R. D. Voyksner, *J. Chromatogr.* **420** (1987) 1.

130. C. G. Edmonds, M. L. Vestal, and J. A. McCloskey, *Nucleic Acids Res.* **13** (1985) 8197.

131. D. W. Phillipson, C. G. Edmonds, P. F. Crain, D. L. Smith, D. R. Davis, and J. A. McCloskey, *J. Biol. Chem.* **262** (1987) 3462.

132. R. S. Annan, G. M. Kresbach, R. W. Giese, and P. Vouros, *J. Chromatogr.* **465** (1989) 285.

133. D. S. Millington, D. L. Norwood, N. Kodo, C. R. Roe, and F. Inoue, *Anal. Biochem.* **180** (1990) 331.

134. D. L. Norwood, C. A. Bus, and D. S. Milligton, *J. Chromatogr.* **527** (1990) 289.

135. C. N. Kenyon, A. Melera, and F. Erni, *J. Anal. Toxicol.* **5** (1981) 216.

136. T. R. Covey, J. R. Crowther, E. A. Dewey, and J. D. Henion, *Anal. Chem.* **57** (1985) 474.

137. F. F. Hsu, C. G. Edmonds, and J. A. McCloskey, *Anal. Lett.* **19** (1986) 1259.

138. R. C. Simpson, C. C. Fenselau, M. R. Hardy, R. R. Townsend, Y. C. Lee, and R. J. Cotter, *Anal. Chem.* **62** (1990) 248.

139. P. Boulenguer, Y. Leroy, J. M. Alonso, J. Montreuil, G. Ricart, C. Colbert, D. Duquet, C. Dewaele, and B. Fournet, *Anal. Biochem.* **168** (1988) 164.

140. H. Y. Kim, J. A. Yergey, and N. Salem, *J. Chromatogr.* **395** (1987) 155.

141. J. Abian and E. Gelpi, *J. Chromatogr.* **394** (1987) 147.

142. R. Richmond, S. R. Clarke, D. Watson, C. G. Chappell, C. T. Dollery, and G. W. Taylor, *Biochim. Biophys. Acta* **881** (1986) 159.

143. M. Guichardant, M. Lagarde, M. Lesieur, and F. De Maack, *J. Chromatogr.* **425** (1988) 25.

144. T. Mizuno, K. Matsuura, and K. Azuma, *Mass Spectrosc.* **34** (1986) 235.

145. G. W. Taylor and D. Watson, *J. Chromatogr.* **394** (1987) 135.

146. H. Y. Kim, J. A. Yergey, and N. Salem, Jr., *J. Chromatogr.* **394** (1987) 155.

147. M. Suzuki, M. Sekine, T. Yamakawa, and A. Suzuki, *J. Biochem.* **105** (1989) 829.

148. M. Suzuki, T. Yamakawa, and A. Suzuki, *J. Biochem.* **108** (1990) 92.

149. A. I. Mallet and K. Rollins, *Biomed. Environ. Mass Spectrom.* **13** (1986) 541.

150. H. Y. Kim and N. Salem, Jr., *Anal. Chem.* **59** (1987) 722.

151. F.-F. Hsu, H. D. Goldman, and W. R. Sherman, *Biomed. Environ. Mass Spectrom.* **19** (1990) 597.

152. D. Watson, G. W. Taylor, and S. Murray, *Biomed. Environ. Mass Spectrom.* **12** (1985) 610.

153. D. Watson, S. Murray, and G. W. Taylor, *Biochem. Soc. Trans.* **13** (1985) 1224.

154. D. J. Liberato, A. L. Yergey, N. Esteban, C. E. Gomez-Sanchez, and C. H. L. Shackleton, *J. Ster. Biochem.* **27** (1987) 62.

155. D. Watson, G. W. Taylor, and S. Murray, *Biomed. Environ. Mass Spectrom.* **13** (1986) 65.

156. C. Eckers, N. J. Haskins, and T. Large, *Biomed. Environ. Mass Spectrom.* **18** (1989) 702.

157. S. Gaskell, K. Rollins, R. W. Smith, and C. E. Parker, *Biomed. Environ. Mass Spectrom.* **14** (1987) 717.

158. N. V. Esteban, A. L. Yergey, D. J. Liberato, T. Loughlin, and D. L. Loriaux, *Biomed. Environ. Mass Spectrom.* **15** (1988) 603.

159. L. O. G. Weidolf, E. D. Lee, and J. D. Henion, *Biomed. Environ. Mass Spectrom.* **15** (1988) 283.

160. Y. Ito, T. Takeuchi, D. Ishii, M. Goto, and T. Mizuno, *J. Chromatogr.* **358** (1986) 201.

161. K. D. R. Setchell and C. H. Vestal, *J. Lipid Res.* **30** (1989) 1459.

9

High-Performance Liquid Chromatography Proton Nuclear Magnetic Resonance On-Line Coupling

Klaus Albert and Ernst Bayer

Institut für Organische Chemie
Auf der Morgenstelle 18
D-7400 Tübingen, Germany

9.1. Introduction

Substantial progress in the field of detectors suitable for high-performance liquid chromatography (HPLC) has been made in the past. Variable-wavelength ultraviolet (UV) detectors have been especially useful in yielding more information about UV characteristics of a compound. Yet, despite all the advances in UV detector performance, there is still a need for a universal detector, especially in cases of compounds without any chromophoric system. In principle, spectroscopic techniques such as mass spectroscopy (MS), infrared (IR), or nuclear magnetic resonance (NMR) spectroscopy can be used as more structural relevant detection techniques. Therefore, strong efforts are made to combine HPLC with IR, MS, and NMR spectroscopy directly.[1,2] In the case of structural recognition of unknown compounds, NMR spectroscopy is one of the most powerful techniques for deriving structural information. The superior stereochemical information content of NMR justifies the current efforts of using an NMR flow cell as a universal detector in HPLC.[2-4]

Several groups are engaged in solving the problems encountered with direct HPLC-NMR coupling and in developing interfaces that fit the needs of HPLC as well as NMR. One of the advantages of using NMR in combination with HPLC is that both HPLC and NMR are conducted in solution, and no evaporation and heating procedures are required. Therefore, the investigation of air- and UV-sensitive compounds is possible. One of the serious disadvantages of the NMR detection principle is the low sensitivity of NMR, leading to larger detection volumes. Because the introduction of extra column dispersion effects leads to a degradation of HPLC separation, the NMR flow cell volume should be as small as possible. Therefore, the development of a continuous-flow NMR detection probe will be a compromise between the needs of HPLC and NMR.

9.2. Effect of Flow on the NMR Signal

In flowing liquids nuclei within a distinct volume (e.g., the detection volume of the NMR measuring coil) exhibit a defined residence time τ. At a constant flow rate complete exchange of all nuclei in the detection volume occurs after the period τ (Figure 9.1). The residence time τ is defined by the ratio between the detector volume V_d and the flow rate u:

$$\tau = V_d/u. \tag{9.1}$$

A shorter residence time τ within the NMR measuring coil results in a reduction of the effective residence time of the particular spin states. Thus, the effective relaxation rates $1/T_n$ are increased by $1/\tau$:

$$1/T_{n \text{ effective}} = \frac{1}{T_i} + \frac{1}{\tau}. \tag{9.2}$$

The influence of the flow rate can be described as a relaxation phenomenon. Due to the limited residence time τ of a nucleus in the NMR detection volume, both T_1, the spin–lattice, and T_2, the spin–spin relaxation times, are reduced in the flowing mode:[5,6]

$$1/T_{1 \text{ flow}} = 1/T_{1 \text{ static}} + 1/\tau \tag{9.3}$$

$$1/T_{2 \text{ flow}} = 1/T_{2 \text{ static}} + 1/\tau \tag{9.4}$$

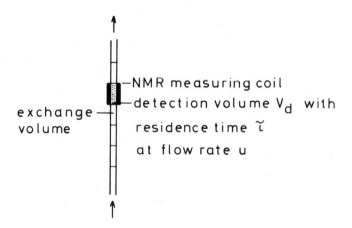

Figure 9.1.   Continuous-flow NMR detection principle.

According to the Heisenberg uncertainty equation, a reduction of the lifetime τ leads to a greater uncertainty in energy and therefore to broader NMR signals. The NMR signal halfwidth W is the reciprocal of the spin–spin relaxation time T_2:

$$W = 1/\pi T_2 \tag{9.5}$$

$$W_{flow} = W_{stationary} + 1/\tau. \tag{9.6}$$

According to Eq. (9.3), an increase in flow rate results in a reduction of T_1. An increase in relaxation rates should result in a higher population of the ground state and thus lead to an increase in signal intensity.

The determination of spin-lattice-relaxation times T_1 can be performed by the inversion-recovery sequence:

Relaxation delay– $180°$ – t – $90°$ –acquisition.

The relaxation delay is necessary to build up complete longitudinal magnetization. After inverting the magnetization by a $180°$ pulse, the longitudinal magnetization reequilibrates. Depending on time t, a $90°$ pulse transfers the partially relaxed longitudinal magnetization to transversal magnetization, giving rise to a free induction decay (FID).

Figure 9.2 shows the magnetization curves of the CH_2 protons of ethanol at three

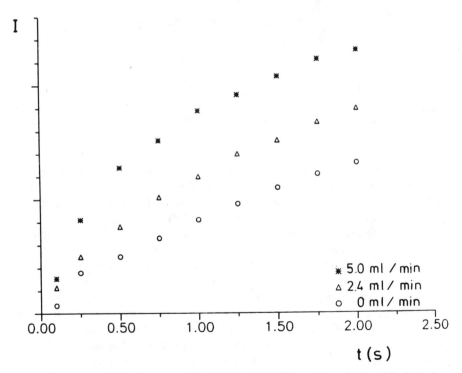

Figure 9.2. Magnetization curves (400 MHz) of the CH_2 protons of ethanol in deuterium oxide at flow rates of 0 (o), 2.4 (Δ), and 5.0 mL/min (*). Arbitrary intensity scale.

stop flow   continuous flow

(2.4 ml/min)

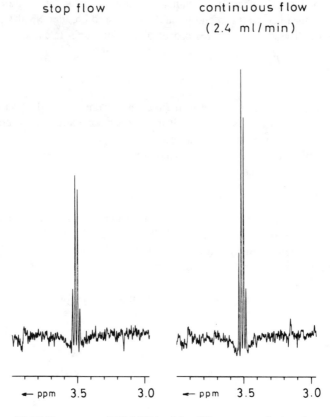

<table>
<tr><td>← ppm</td><td>3.5</td><td>3.0</td><td>← ppm</td><td>3.5</td><td>3.0</td></tr>
</table>

Figure 9.3.   ¹H NMR spectrum (400 MHz) of the CH₂ protons of ethanol recorded under stopped-flow and continuous-flow (2.4 mL/min) conditions.

different flow rates using a detection volume of 188 μL. The data are obtained by an inversion-recovery sequence using a prepolarization volume of 2 mL to obtain a Boltzmann equilibrium of the nuclei before entering the flow cell. The prepolarization volume was built up by coiling two meters of PTFE tubing (1.1 mm i.d.) near the probe bottom in the region of a strong magnetic field. This volume is equivalent to the solvent volume in an HPLC column of 4.0 × 250 mm. The faster reequilibration of magnetization at flow rates of 2.4 and 5.0 mL/min can be seen in different magnetization curves (Figure 9.2). The different slopes of magnetization in the stationary mode and at the flow rates 2.4 and 5.0 mL/min are defined by different relaxation times of 3.4(3), 2.0(3), and 1.0(6) s, respectively. Thus, the intensity of the NMR signal increases with increasing flow rate if a Boltzmann distribution is obtained prior to NMR detection (Figure 9.3). The enhancement rate in comparison to stationary measurement is dependent on the detector-volume–to–flow-rate ratio and on the acquisition parameters such as flip angle and pulse repetition time.

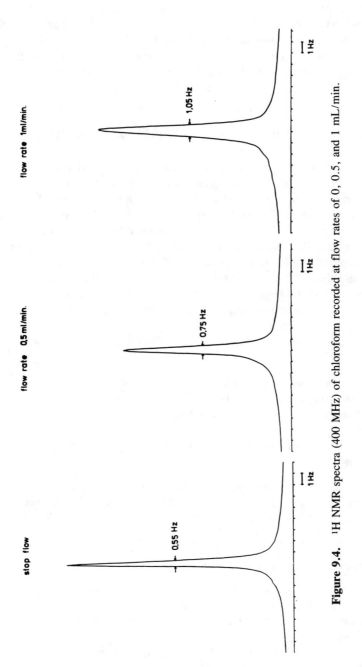

Figure 9.4. ^{1}H NMR spectra (400 MHz) of chloroform recorded at flow rates of 0, 0.5, and 1 mL/min.

The drawback of continuous flow NMR is the influence on T_2 according to Eq. (9.4). An increasing flow rate should result in an increased signal linewidth. Figure 9.4 shows the signal of chloroform, recorded at different flow rates using a detection volume of 44 μL. The static signal linewidth is 0.55 Hz; at a flow rate of 0.5 mL/min it is increased to 0.75 Hz, and at a flow rate of 1.0 mL/min it exceeds 1.0 Hz. The experimentally recorded line-broadening values are in good agreement with predicted values according to Eq. (9.6).

9.3. Development of Continuous-Flow NMR Probes

Several authors have described measurement arrangements that allow the registration of stopped-flow and continuous-flow kinetics in the continuous-wave and in the pulse Fourier transform (PFT) mode.[5-33] The first application in the field of HPLC-NMR coupling was performed by Watanabe and Niki.[34] They used a conventional NMR tube to record PFT NMR spectra of separated HPLC peaks with the stopped-flow technique. Because of the great dead volumes encountered with this experimental approach, different groups tried to develop special continuous-flow NMR probes suitable for continuous-flow and stopped-flow HPLC-NMR coupling.[34-54] To overcome the problems of low concentration, all groups used the pulse Fourier transform technique to accumulate signals during the passage of the peaks through the NMR detection region.

Four main problems must be solved in the development of continuous-flow NMR probes:

1. The geometry of the detection cell should enable laminary flow;
2. Wall (memory) effects must be avoided;
3. At a distinct detection volume the NMR sensitivity should be as large as possible;
4. The obtained NMR resolution should allow the interpretation of complex spin systems;

The volume of the measuring cell is dependent upon the NMR sensitivity of the measuring assembly. The signal-to-noise (S/N) ratio of an NMR detection cell is defined by the following parameters[55]:

$$S/N \approx N\gamma I(I+1)B_0^{3/2}\phi(QV_S)^{1/2}\,b^{-1/2}\,T^{-3/2}\,f^{-1}. \tag{9.7}$$

Increasing sensitivity values are obtained by:

1. Increasing the number of nuclei N within the detection volume;
2. Measurement of nuclei with high magnetogyric ratio γ;
3. Measurement of nuclei with high spin quantum number I;
4. Increasing the magnetic field strength B_0;
5. Increase the filling factor ϕ of the NMR coil, $\phi = V_S/V_C$, where V_S is the sample volume and V_C the NMR coil volume;
6. Increasing the quality factor Q of the coil;

7. Increasing the sample volume V_S;
8. Reduction of the receiver bandwidth b;
9. Reduction of the measuring temperature T;
10. Reduction of the noise figure of the preamplifier f.

In the construction of NMR probe heads, points 5–7 play a major part. Measuring coils of the same geometry and of the same composition exhibit the same quality factors Q; therefore, the S/N ratio is dependent upon the sample volume and the filling factor ϕ of the NMR coil. In contrast to conventional NMR probe heads, the detection volume should be as small as possible, a compensation for the resulting smaller sensitivity can only be performed by increasing the filling factor. This is possible by the direct fixation of the NMR coil at the glass of the continuous-flow cell. Thus, the difference between coil and sample volume is the volume of the glass cylinder surrounding the detection area; the filling factor is close to 1. The disadvantage of this approach is that rotation of the NMR tube is impossible and compensation of magnetic field inhomogeneities is more difficult.

The signal linewidth increases, especially at the foot of the resonances. Concerning the same peak area of a signal under rotating and nonrotating conditions, its intensity is reduced in the nonrotating case because of the increasing linewidth. This is the reason why the stationary sensitivity in the NMR flow cell is lower than in the rotating NMR tube despite all efforts at optimizing cell and coil geometry.

A further question is the geometrical position of the NMR flow cell. A vertical fixation of the flow cell together with its geometry may enable laminar flow. If the flow direction is guided against the earth's gravity field, air bubbles, which may have entered the mobile phase, can be easily swept out of the measuring region. Silylation of the inner glass cell leads to an efficient protection against any absorption of eluents.

9.3.1. Determination of Detector Dispersion Effects

Because of the low sensitivity of the NMR detection technique, greater detection volumes are necessary than in the case of, for example, UV detectors. The dispersion effects of several NMR flow cells were checked by directly measuring the resulting peak widths of an HPLC test mixture in the detector cell with the help of a modified fluorescence detector.[48] The fluorescence technique is insensitive to the glass of the measuring cell. The NMR glass cell under investigation was fixed within the beam of a modified fluorescence detector and connected to an HPLC column (Figure 9.5). Two analytical HPLC columns (i.d. 4.0 mm) with lengths of 12.5 cm and 25 cm and a semipreparative column (250 × 8 mm) were used. The test mixture consisted of 4 dansylated amino acids (aspartic acid, glycine, arginine, and alanine) in an isocratic eluent of methanol/water (60/40, pH 2). The retention times t_R together with the resulting HPLC peak widths $W_{1/2}$ yielded the plate numbers N according to

$$N = 5.54(t_R/W_{1/2})^2. \tag{8}$$

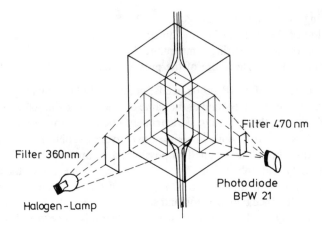

Figure 9.5. Modified fluorescence detection arrangement for the determination of peak broadening in different flow cells.

Plate heights h were calculated according to $h = (L/N)d_p$ (L column length, d_p particle size). The resulting relationship between calculated plate height and capacity factor reveals that plate heights are adversely affected by capacity factors below 2.5. In the case of short analytical columns (12.5 cm length), only the flow cell of 48 μL volume does not cause a substantial increase in plate height (Figure 9.6). In the case of long analytical columns, even a larger detection volume of 110 μL does not lead to a dramatic increase in plate height (Figure 9.7). Even greater detection volumes can be tolerated when semipreparative columns are used (Figure 9.8).

9.3.2. Iron Magnet Systems

Due to the low sensitivity of iron magnet systems, large detection volumes are needed to guarantee a sufficient number of nuclei in the detection area. Whereas Buddrus and Herzog[38–40,46] used a routine NMR tube with a detection volume of about 700 μL, our group[35] developed a ^{1}H-NMR continuous-flow probe with a detection volume of 416 μL (Figure 9.9). Field-frequency stabilization was achieved with a D_2O ampoule (external lock), which was placed near the actual NMR cell (Figure 9.9). PTFE capillary connection tubes (i.d. 0.23 mm) were connected to the inlet and outlet glass capillary by shrink-fit tubing. The obtained resolution was 2.5 Hz (Figure 9.10), which was sufficient for scouting experiments.

9.3.3. Cryomagnet Systems

In cryomagnets the direction of the main magnetic field B_o is perpendicular to that in iron magnets. If a cylindrical measuring coil is positioned in the vertical direction parallel to the magnetic field, the number of interactions with the magnetic flux is reduced, and the sensitivity is reduced. Moreover, the Q factor of the Helmholtz

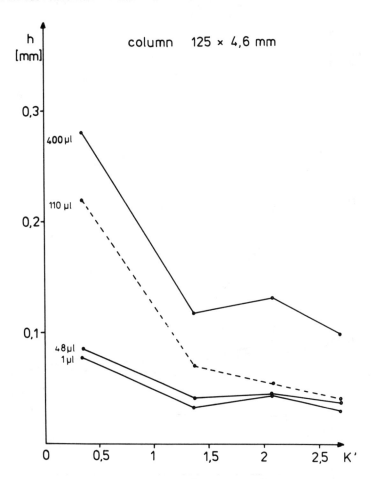

Figure 9.6. Effect on plate height (h) of different flow cell volumes using an analytical column (125 × 4.6 mm).

coil commonly used in a vertical position is lower than that of a horizontally oriented solenoid coil. Therefore, a horizontal fixation of the measuring coil would be desirable. Unfortunately, the shim systems in all commercially available high-resolution NMR instruments are designed for vertical measuring assemblies, because in routine operation NMR tubes are exchanged from the top of the cryomagnet. Hence, at the moment homogenization of a horizontal region of 15 mm with a resolution of better than 1 Hz is impossible. However, an NMR resolution better than 0.5 Hz should be obtained by a vertically positioned, nonrotating flow cell arrangement.

The selection of the detection volume of the NMR flow cell is a compromise between the requirements of HPLC and NMR. According to the theory of Scott and Kucera, the optimal detection volume is 1–5 µL, when an HPLC column of 4 × 250

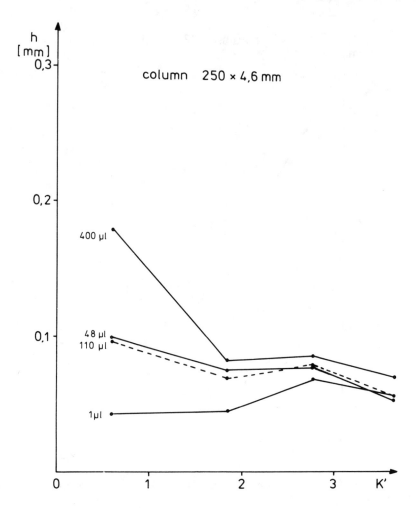

Figure 9.7. Effect on plate height (h) of different flow cell volumes using a long analytical column (250×4.6 mm).

mm is used.[56] In contrast, the detection volume within a routine 5 mm tube is 250 μL. Within this volume, sufficient nuclei are present to guarantee a good signal-to-noise spectra ratio in most routine measurements. Concerning the registration of continuous flow, the dependence of signal linewidth on the residence time τ and the detection volume–to–flow-rate ratio causes further problems.

Table 9.1 shows residence times together with eminent linewidth broadenings of five different flow cell volumes. At flow rates of 1 mL/min, an increase of 0.1 Hz is only obtained with flow cells of a volume of > 100 μL. This volume is apparently larger than that in routine UV detection. To avoid line-broadening effects of more than 1.1 Hz, with detector volumes of 16 and 8 μL, only flow rates of < 0.5

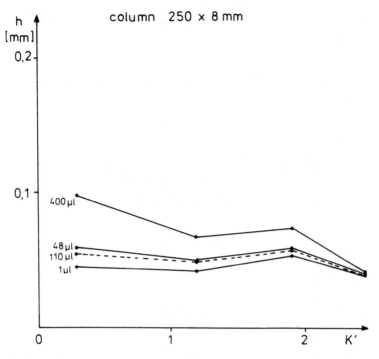

Figure 9.8.   Effect on plate height (h) of different flow cell volumes using a semipreparative column (250 × 8 mm).

Table 9.1.

	Data of different flow cells			
Volume [μl]	Residence time $\tilde{\iota}$ [s]		Linewidth broadening $1/\tilde{\iota}$ [Hz]	
	flow rate 0.5 mL/min	flow rate 1.0 mL/min	flow rate 0.5 ml/min	flow rate 1.0 ml/min
8	0.96	—	1.04	—
16	1.92	—	0.52	—
44	5.28	2.64	0.19	0.38
120	14.40	7.20	0.07	0.14
188.5	22.62	11.31	0.04	0.09

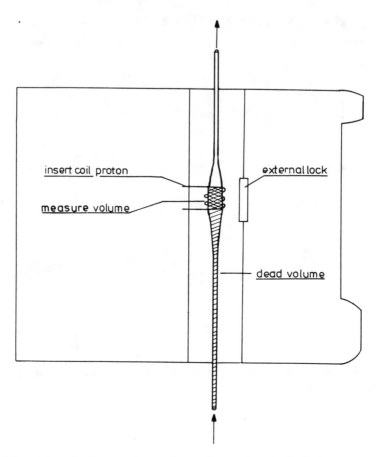

Figure 9.9. Schematic diagram of a continuous-flow probe suitable for iron magnets.

mL/min can be employed. In the construction of high-field NMR probes double saddle Helmholtz coils are mainly used. Therefore, in all approaches for continuous-flow probes suitable for cryomagnets with a resolution below 1 Hz, Helmholtz coils with upward flow direction are applied. Haw and Dorn[41,42] used a 5 mm NMR tube (Figure 9.11), Laude and Wilkins[47] a 1.5 mm capillary (Figure 9.12) and our group[43,54] an inverted U-type detector with 2 mm and 3 mm i.d. (Figure 9.13). Thus, different detector volumes of 20–120 μL could be obtained. Dorn and coworkers inserted a 5 mm tube into an 8 mm NMR tube to which the NMR Helmholtz coil was affixed (Figure 9.11). Gradual tapers were used at the inlet and outlet of the active volume to avoid formation of eddy flow effects. The volume within the coil region is approximately 120 μL. The flow insert was mounted in a probe housing similar to those on commercially available probes.

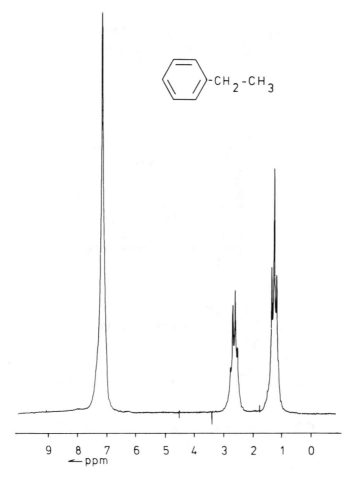

Figure 9.10.   Continuous-flow [^1]H NMR spectrum (90 MHz) of ethylbenzene in CCl_4, flow rate 1 mL/min.

Wilkins and co-workers inserted a 1.5 mm capillary tubing with total observation volumes of 20, 28, and 53 μL into a standard 5 mm proton probe (Figure 9.12).

In cooperation with the Bruker Company, Karlsruhe, Germany, the Tübingen group modified routine NMR probes for use in stopped-flow and continuous-flow detection. The measuring arrangement consists of a vertically fixed nonrotating glass tube with internal diameters of either 2 or 3 mm and tapering at both sides to the outer diameter of the inlet and outlet PTFE tubings (Figure 9.13). These tubings are fixed to the glass cell by shrink-fit tubings. The NMR Helmholtz detection coil is directly fixed to the glass cell, resulting in detector volumes of 44, 56, and 120 μL. The total measurement arrangement is located within a glass dewar, with a

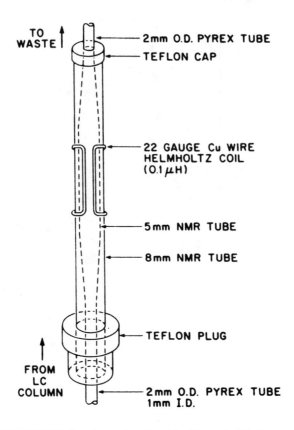

Figure 9.11.   LC-^{1}H NMR flow insert developed by Dorn et al.[41,42]

thermocouple inserted which facilitates low- and high-temperature measurements under well-defined conditions.

The sensitivity (in the stationary mode) of this flow cell development approaches that of routine probe heads to about 60–70%, whereas the resulting stationary resolution is between 0.2 and 0.5 Hz. An important feature of ^{1}H-NMR probes is the signal linewidth at the height of the ^{13}C satellites. This (hump test) is usually recorded with a sample of chloroform in acetone-d_6. Figure 9.14 shows the continuous-flow spectrum (300 MHz) of chloroform recorded at a flow rate of 1 mL/min at a 120 μL flow cell. The linewidth at the height of the ^{13}C satellites is 14 Hz; at $\frac{1}{5}$ of this altitude it is 30 Hz. With a 56 μL flow cell at a flow rate of 0.5 mL/min, a linewidth of 12/24 Hz at 0.5/0.1% intensity of the CHCl$_3$ signal can be obtained.[57] These values are approximately the same as those of conventional probes with rotation of a 10 mm NMR tube.

With the present state of development of NMR Helmholtz coils, a detection volume of about 40–120 μL is necessary to obtain continuous-flow ^{1}H NMR spectra with appropriate S/N values with a time resolution lower than 10 s. Flow

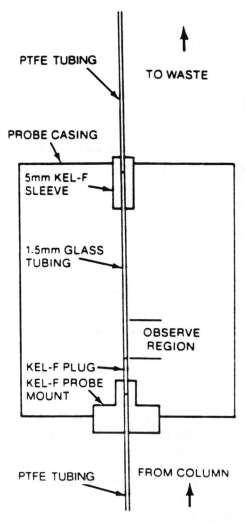

Figure 9.12. Schematic diagram of the flow cell used by Wilkins et al.[47]

cells with volumes < 60 μL are used with a flow rate of 0.5 mL/min, whereas at cell volumes > 100 μL more rapid HPLC separations can be performed with a flow rate of 1 mL/min.

9.4. Experimental Arrangement for HPLC-NMR Coupling Using Cryomagnets

The experimental arrangement used in HPLC-NMR coupling differs from group to group. Dorn housed the HPLC hardware in a Plexiglass cabinet located approximately 50 cm from the outer edge of the magnet dewar.[41,42] A refractive index (RI)

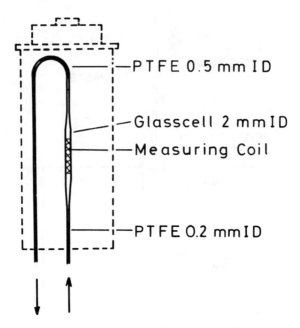

Figure 9.13. Schematic diagram of the continuous-flow cell measurement arrangement developed by the Tübingen group.[43,54]

detector was used to obtain classical chromatograms. The NMR flow cell was connected directly to the outlet of the RI detector via a length of stainless-steel capillary tubing. A Jeol FX-200 nuclear magnetic resonance spectrometer equipped with an Oxford 4.7-T superconducting solenoid magnet (54 mm bore) was used to obtain ^1H spectra at 199.5 MHz.

Wilkins et al. situated the HPLC column within the room-temperature bore of a Nicolet 300 MHz wide-bore magnet (Figure 9.15), and the connection to the pump was made with stainless-steel or PTFE tubing.[47] Three meters of tubing was fixed between the pump and the injection valve.

The Tübingen group in cooperation with the Bruker company used their continuous-flow probe heads together with standard narrow-bore superconducting magnets (Bruker AC 300, Bruker WM 400, Bruker AM 500).[43,48,53,54,57] The changeover between standard and continuous-flow probe head can easily be performed within a few seconds by loosening and tightening two screws (Figure 9.16). The readjustment of magnet field homogeneity can be performed in 15–30 min by either shimming at the lock signal of acetone-d_6 or on the ^1H NMR interferogram of chloroform in acetone-d_6. The HPLC column (4.0 × 250 mm) is fixed at the bottom at the cryomagnet and attached to the continuous-flow probe by a 10 cm stainless-steel capillary (0.25 mm i.d.). Thus, the transfer volumes between the HPLC column and the detection region are 33 μL (44 μL flow cell) and 110 μL (120 μL flow cell). The injection system is fixed by brass connectors to one stand of the cryomagnet

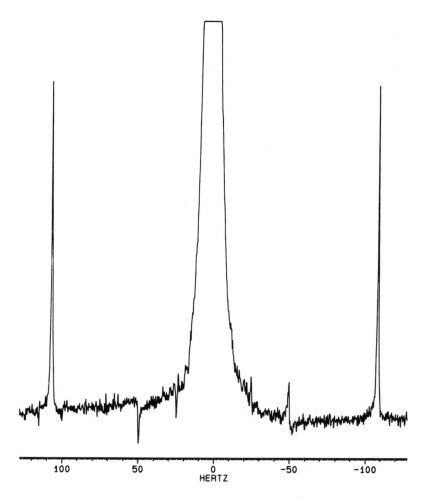

Figure 9.14. ¹H NMR signal line shape of chloroform in acetone-d_6 (hump test), measured in a 120 μL continuous-flow probe (300 MHz) at a flow rate of 1 mL/min.

and connected to the HPLC column by 20 cm of stainless-steel capillary. The HPLC pump together with the solvent reservoir is located at least 1 m from the magnet. With this experimental arrangement, magnet homogeneity shimming can be readily performed, and HPLC columns can be changed with any disturbance of magnet field homogeneity.

9.5. Applications in Absorption Chromatography

Usually in high-resolution NMR, deuterated solvents are employed, the deuterium signals of which are used for field-frequency stabilization. Because most deuterated

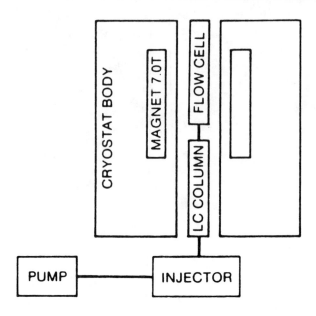

Figure 9.15. Block diagram of an on-line HPLC-NMR system with "in bore" design.

solvents in HPLC-NMR coupling are too expensive, only halogenated eluents and a few deuterated solvents such as $CDCl_3$, D_2O, and CD_3OD are economically feasible. When halogenated solvents such as CCl_4 were used, field-frequency stabilization could be achieved with the external lock arrangement (see Section 9.3.2).

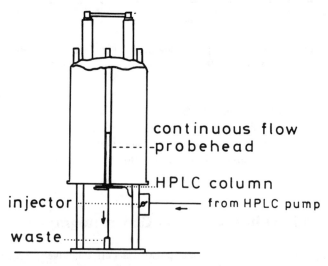

Figure 9.16. Experimental arrangement for on-line HPLC-NMR coupling with easy access to HPLC column.

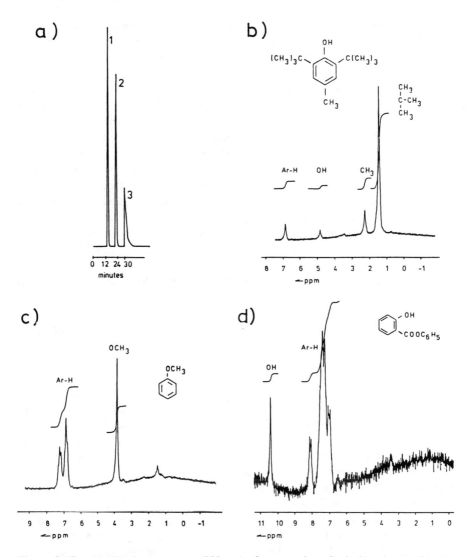

Figure 9.17. (a) UV chromatogram (280 nm) of a separation of substituted aromatics: 1, ionol; 2, anisole; and 3, salol. (b) Continuous-flow NMR spectrum (90 MHz) of ionol. (c) Continuous-flow NMR spectrum (90 MHz) of anisole. (d) Continuous-flow NMR spectrum (90 MHz) of salol.

9.5.1. Substituted Aromatics

The first continuous-flow HPLC-^{1}H NMR experiment was performed in the non-protonated solvent carbon tetrachloride at a 90 MHz iron magnet.[35] The HPLC instrument was located 1 m from the magnet, and the outlet of the UV detector was connected to the continuous-flow probe (shown in Figure 9.9) by PTFE tubing (i.d. 0.25 mm).

The substituted aromatics ionol, anisole, and salol could be separated at a silica gel column using carbon tetrachloride as eluent. The solvent was first dried over Al_2O_3 and saturated with D_2O. The UV chromatogram showed a complete resolution of the individual components [Figure 17(a)].

After the injection of 10 μL of an equimolar mixture ((50 μmol) of ionol, anisole, and salol in carbon tetrachloride, the separation was carried out with a flow rate of 1 mL/min within 26 min. Every 51 s after the occurrence of the peak maximum in the UV detector, the NMR measurement was started. Thus, different numbers of scans, depending upon the respective peak width, were made. For a sweep width of 3000 Hz and the applied memory of 4 K, the acquisition time for a particular scan is 0.7 s. The interferograms of the three separated components were stored in different memories, and evaluated after the separation. The first peak (ionol) was observed to have an elution time of 144 s, and the measuring time for its ^{1}H-NMR spectrum amounted to 178 scans, equivalent to 121 s [Figure 17(b)]. Analogously, the elution time for anisole was 220 s [297 scans, 202 s, Figure 17(c)], and for salol it was more than 800 s [744 scans, 526 s, Figure 17(d)]. The complete interferograms of the three components were run within their elution time. In spite of the greater linewidth of the ^{1}H-NMR spectra, a classification is possible based upon the obtained data (ppm values, integration ratio). The detection of the intramolecular hydrogen bond formation in the ^{1}H-NMR spectrum of the third peak [Figure 17(d)] demonstrates the stereochemical information content of the method.

9.5.2. Hydrocarbons

Using the cell development shown in Figure 9.11, a model mixture of an artificial fuel sample (n-butylbenzene, m-xylene, tetralin, naphthalene, dodecane, isooctane, n-hexane, n-nonane, n-hexadecane, and n-pentane) was separated in a mixture of 1,1,2-trichlorotrifluoroethane (Freon 113) and $CDCl_3$ (95/5 v/v) using a silica gel packing that was derivatized to introduce amino and cyano functionalities (silica gel-PAC column), 250 × 9 mm i.d.).[41] NMR spectra were obtained at a flow rate of 1 mL/min, and the first 16 spectra were taken across the aliphatic peak region over a period of 15 s each (Figure 9.18). Long-chain n-alkanes elute early, whereas the branched alkane (isooctane, detected by the methine signal) elutes in spectra 5–8. Spectrum 10 shows almost exclusively hexane and pentane. Most of the alkane material is present in spectra 5–7, representing a volume of 750 μL. This volume compares favorably with the width of the RI trace alkane peak, demonstrating that extracolumn band broadening is not excessive with the employed scale of chromatography (e.g., a 9 mm i.d. column). (This find is consistent with observed dispersion effects of 120 μL NMR flow cells used together with semipreparative columns.) n-Butylbenzene is present in spectra 22–24. The signal-to-noise ratio is quite good, at the expense of some resolution. Although baseline chromatographic resolution is not achieved between m-xylene and tetralin, the resonances of their aliphatic protons are clearly resolved along the chemical shift axis.

The measured quantities of the four aromatic compounds, determined by integration, give good agreement with the known values. The amount of naphthalene in

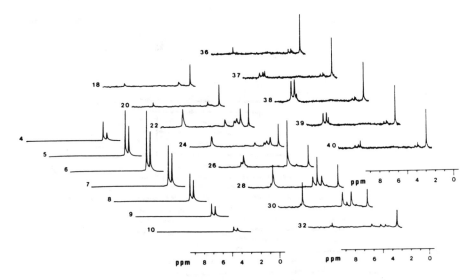

Figure 9.18. LC-^{1}H NMR profile (200 MHz) for model mixture of an artificial fuel sample.

spectrum 40 was determined from the integrations of files 37–40 and the known quantity injected. This value is approximately 25 µg and suggests that the limit of detection is about 10 µg, depending on the sample and conditions used. Also visible in these spectra are signals due to residual chloroform and a trace of water stripping off the column.

9.6. Applications in Reversed-Phase Chromatography

Most of HPLC separations are performed in the reversed-phase mode using protonated solvents. Therefore, the realization of a reversed-phase experiment was an important step.[43]

9.6.1. Separation of Acetonitrile

Figure 9.19 shows the ^{1}H NMR chromatogram (400 MHz) of an HPLC separation (C_{18} column) of two plastifiers (2.22 and 4.18 mg each) in acetonitrile recorded at a flow rate of 0.5 mL/min. 36 scans were co-added for each spectrum, resulting in a time resolution of 30 s.

This example shows that an immediate classification of the compounds investigated can be performed by continuous-flow ^{1}H NMR detection. The existence of aliphatic, ester, and aromatic groups can be readily seen at the distinct chemical shift values.

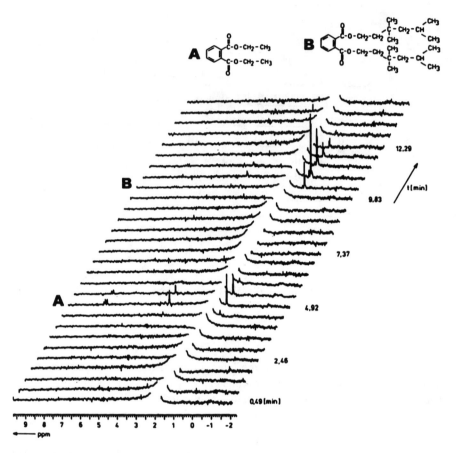

Figure 9.19. ¹H NMR chromatogram (400 MHz) of a reversed-phase separation of two plastifiers.

9.6.2. Techniques of Solvent Signal Suppression in Flowing Liquids

In a reversed-phase separation with two or more solvent signals, solvent suppression has to be performed prior to ¹H NMR detection to solve dynamic range problems. The use of solvent presaturation schemes with only one conventional detection coil in the continuous-flow probe is not very favorable because presaturated nuclei are being continuously removed from the NMR flow cell. Up to now two different techniques have been applied to overcome the problem of protonated solvents. The first is the use of a selective excitation technique to excite only a distinct region of the NMR spectrum.[58] Here the signals of protonated solvents do not contribute to the interferogram; therefore, dynamic range problems are negligible.

The second possibility is to use a binomial solvent suppression technique to

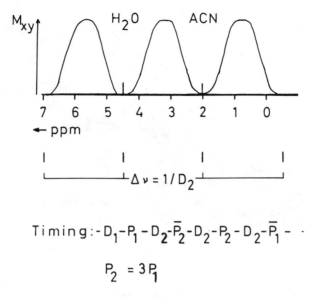

Figure 9.20.   Transverse magnetization excited by the $1\bar{3}3\bar{1}$ sequence as a function of the offset setting in acetonitrile–water mixtures. ACN, acetonitrile.

perform signal intensity reduction in the continuous-flow mode. The most commonly applied techniques are the 11 and $1\bar{3}3\bar{1}$ sequences.[59-62] Both techniques are able to suppress two solvent signals simultaneously by at least two orders of magnitude. In contrast to presaturation techniques, no additional time delay is needed.

Zero spectral density at the solvent resonances is generated by a series of pulses with appropriate delays, based upon the nth series of binomial coefficients. There exist two sets of binomial solvent suppression techniques: with and without 180° phase alternations of transmitter pulses, resulting in a sine and cosine scheme of solvent nonexcitation. The 11, 121, 1331, 14641 set without phase alternations results in a sine nonexcitation scheme; the $1\bar{1}$, $1\bar{2}1$, $1\bar{3}3\bar{1}$, $14\bar{6}4\bar{1}$ in the opposite.

The 1-1 hard pulse, simplest of the binomial suppression methods, consists of a $P_1-D_2-P_1$ sequence, with $2P_1$ equal to a 90° pulse. If the pulse carrier is spaced $\Delta\nu$ from the solvent resonance, then following a 45° pulse, a delay $D_2 = 1/2\ \Delta\nu$ allows precession of the solvent magnetization at exactly 180°. Application of an equivalent 45° pulse nulls the solvent magnetization along the z axis. All other resonances not equal to $(2n + 1)\Delta\nu$ retain transverse magnetization and are detected. This sequence can be used to suppress the water signal.

Solvent nonexcitation of two signals such as water and acetonitrile can be performed with the $1\bar{3}3\bar{1}$ sequence (Figure 9.20). Transverse magnetization is a function of the offset setting. If the pulse carrier frequency is adjusted to the water resonance, further nulls of transverse magnetization are generated at intervals $\Delta\nu = 1/D_2$. With a 300 MHz instrument, a frequency difference of 965 Hz (3.22 ppm) between the water and the acetonitrile signals results in a delay D_2 of 1 ms.

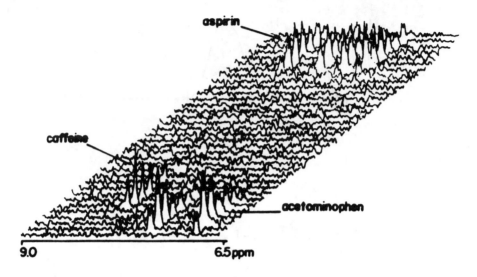

Figure 9.21. Stacked plot of the aromatic region of ^{1}H NMR spectra (300 MHz) from the HPLC/NMR separation of a three-compound mixture of analgesics.

9.6.2.1. Analgesics

A three-component mixture consisting of 200 μg of acetaminophen, 400 μg of caffeine, and 500 μg of acetylsalicylic acid was separated in 50:50 H_2O/acetonitrile (with H_2O buffered at pH 4.4) at a flow rate of 0.5 mL/min using a 10 μm 30 cm × 4.6 mm C_8 bonded phase column (Figure 9.21).[51] The 1-1 hard pulse procedure was used to suppress both resonances of H_2O and acetonitrile in a 28 μL flow cell.[51] A spectrum is a file of four co-added scans stored every 2.3 s.

9.6.2.2. Substituted Aromatics

Figure 9.22 shows the contour plot (500 MHz) of a separation of 7 aromatic compounds (70 μg each) in acetonitrile/water (50:50) at a flow rate of 1 mL/min.[53] Here, solvent suppression of the acetonitrile and the water resonance was performed by application of the $1\bar{3}3\bar{1}$ sequence. In the contour plot only the chemical shift range between 2.6 and 10.3 ppm is plotted against the separation times between 0 and 40 min. Therefore, only the residual water resonance is seen at 4.4 ppm throughout the whole separation. At a retention time of 8 min, the signals of phenol protons appear, followed by the signals of benzaldehyde at t_R = 12 min and of acetophenone at t_r = 13 min. The observed chemical shifts in these compounds clearly indicate the existence of an aldehyde and an alkoxy or methoxy substituent. The following signals of nitrobenzene at t_R = 23 min are very weak, whereas the signal of methylbenzoate at t_R = 25 min can clearly be detected. The signals of the last two compounds are due to anisole (t_R = 30 min) and to benzene (t_R = 34 min).

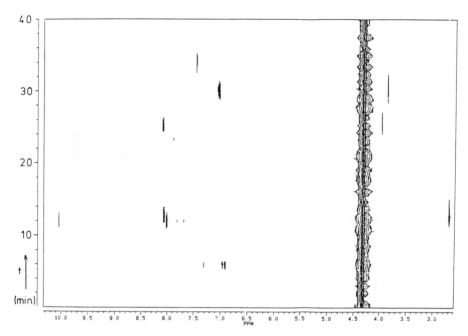

Figure 9.22.   Contour plot of the ^{1}H NMR chromatogram (500 MHz) from the HPLC/NMR separation of a seven-compound mixture of substituted aromatics.

With the exception of nitrobenzene, the ^{1}H NMR peaks in this chromatogram reveal more information than any diode array detection.

9.6.2.3. Applications with Selective Excitation

If only a small spectral region of the ^{1}H NMR spectrum provides sufficient information for peak identification, the selective excitation technique can be used. Frequency-selective excitation may be performed by application of a Gaussian-shaped radio frequency pulse that, contrary to a rectangular pulse, does not produce excitation lobes. The pulse width for excitation of a bandwidth of about 700 Hz of a spectral width of 4800 Hz (300 MHz instrument) needs a 4 ms "soft" pulse, which may be generated by a selective excitation unit.

9.6.2.3.1. Cyclopropyl-Containing Compounds

The metabolytes of 2-(dicyclopropylmethyl)amino-Δ^2-oxazoline dihydrogenophosphate (Figure 9.23 shows the ^{1}H NMR spectrum) were found all to contain the cyclopropyl group, and this group was proposed as a detection tag, indicating degradation products of the drug.[54] Because of the characteristic high-field shift (0.2–1.2 ppm) of the cyclopropyl proton signals in the ^{1}H NMR spectrum, on-line HPLC/^{1}H NMR coupling would appear the ideal means of investigation. By adjust-

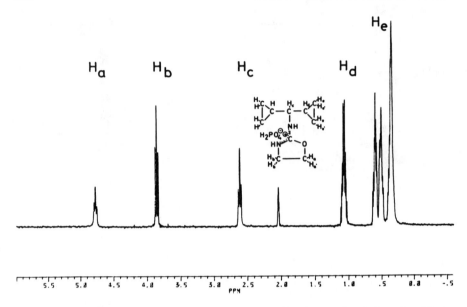

Figure 9.23. ¹H NMR spectrum (400 MHz) of 2-(dicyclopropylmethyl) amino-Δ²-oxazoline dihydrogenophosphate.

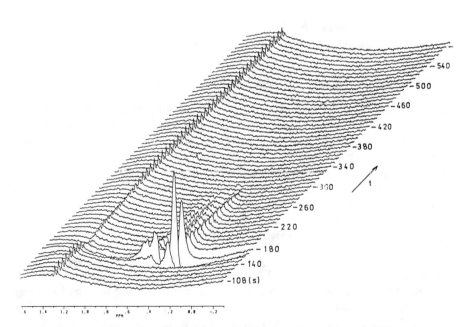

Figure 9.24. Stacked plot of aliphatic region of ¹H NMR spectra (300 MHz) from the HPLC/NMR separation of two cyclopropyl-containing compounds [2-(dicyclopropylmethyl)amino-Δ²-oxazoline dihydrogenophosphate and dicyclopropyl ketone].

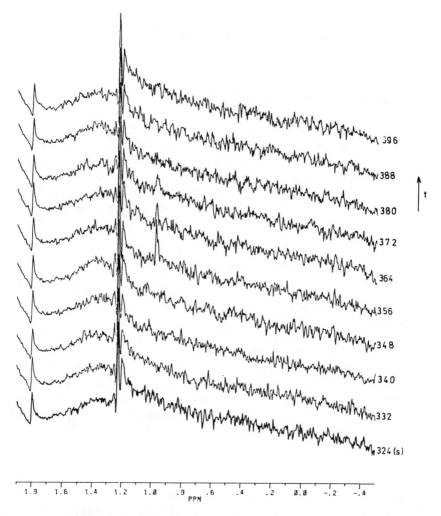

Figure 9.25. Expended ¹H NMR chromatogram of Figure 9.23 in the time interval 324–396 s.

ing the pulse carrier in the middle of the region of interest, only the ¹H chemical shift range between 0.7 and 1.7 ppm was investigated by use of a Gaussian-shaped pulse. Figure 9.24 shows the ¹H NMR chromatogram (stacked plot) of a separation of 4 mg of oxazoline drug and 120 μg of dicyclopropyl ketone. A 250 × 4 mm C_8 column was used at a flow rate of 1 mL/min with a solution of 0.14% triethylamine in acetonitrile–water (37.5:62.5) buffered to pH 3 by H_3PO_4. Throughout the whole separation, the signal of the triethylamine methyl protons at 1.25 ppm is seen, while the cyclopropyl protons of the oxazoline compound appear at $t_R = 180$ s. In addition, a small signal at $t_R = 356$ s can be detected. Upon expanding the plot over the time of interest (Figure 9.25), it is clear that the resonances at 1.0 ppm are from the H_b and H_b' protons of dicyclopropylketone.

According to this separation, the present detection limit of compounds that are structurally related to dicyclopropylketone is better than 3% in a 120 μL flow cell at 7.0 T.

9.7. Stopped-Flow Technique

In structure elucidation of unknown compounds, the application of two-dimensional assignment techniques is very helpful. Thus, homo- and heteronuclear connectivities can easily be detected. Because the diffusion constant of water is very small, HPLC separations can be stopped resulting in small dispersion effects.[56,63] Therefore, it is possible in an on-line HPLC-NMR experiment, after determination of the maximum peak concentration in the NMR flow cell, to stop the flow and to record a 2D correlated spectrum.

9.7.1. Cyclopropyl-Containing Drug

Figure 9.26 shows the contour plot of a homonuclear shift correlated experiment (COSY 90) of the aliphatic region of 2-(dicyclopropylmethyl)amino-Δ^2-oxazoline dihydrogenophosphate.[54] The 2D spectrum was recorded within 23 min in acetonitrile/water after stopping the flow. Suppression of the solvent signals was performed by application of the $1\bar{3}\bar{3}1$ sequence. Even the intensity of the cross peaks at 0.35 ppm is reduced in the F_2 axis; in the F_1 axis the correlation between the H_d protons at 1.05 ppm together with the set of H_c and H_c protons at 0.32, 0.50, and 0.59 ppm is clearly seen. This example does not show a fully optimized 2D correlation but indicates that it is possible to perform 2D assignment techniques in on-line HPLC-NMR coupling.

9.7.2. Polyester

Another application of a homonuclear shift correlated experiment (phase-sensitive COSY) is shown in Figure 9.27. Here, an HPLC peak of a polyester separation in CD_3OD and D_2O was examined using a 56 μL flow cell.[57]

Both stopped-flow applications show that on-line HPLC-NMR coupling will be a great help in the sometimes laborious task of the isolation and structural assignment of naturally occurring compounds as well in the structural elucidation of drug metabolites.

9.8. Further Developments in HPLC-NMR Coupling

Theoretically, in the continuous-flow mode, a further improvement in sensitivity could be obtained if instead of the Helmholtz coil used in vertical measurement (Figures 9.11–9.13), a horizontally located solenoid coil or toroidal coil could be used.[55] Dorn and co-workers performed experiments with toroidal coils.[64,65] They were able to raise sensitivity values by a factor of two, but the resulting signal line

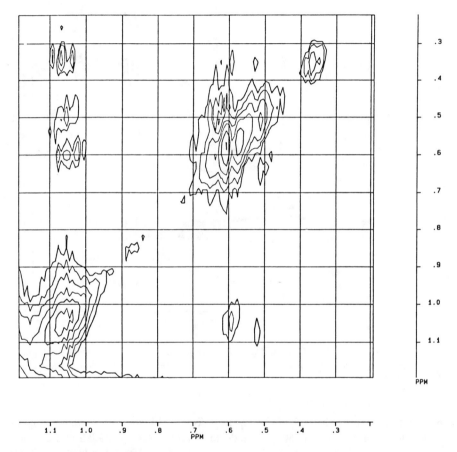

Figure 9.26. Contour plot (300 MHz) of a ¹H homonuclear shift correlated experiment (COSY 90) of a cyclopropyl-containinng drug. Stopped-flow measurement in protonated solvents.

shape was in the 15 Hz range. This resolution is not sufficient for high-resolution NMR and has to be improved to the range of 2–3 Hz. A better resolution may be obtained by a shim system that is optimized for magnetic field homogenization in the horizontal direction. Then detector volumes of some microliters can be applied, which allows the use of microbore columns and deuterated solvents.

With conventional shim systems, optimal NMR resolution was independently

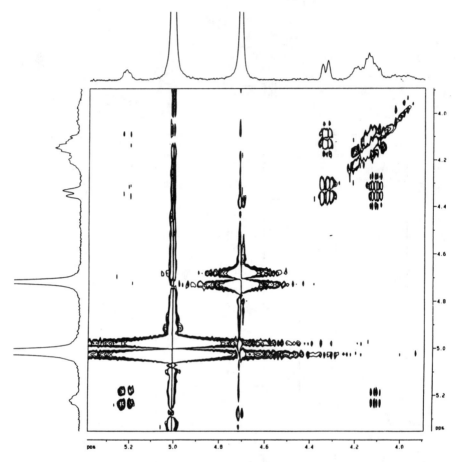

Figure 9.27. Contour plot (500 MHz) of a ¹H homonuclear shift correlated experiment (phase-sensitive COSY) of a polyester. Stopped-flow measurement in deuterated solvents.

obtained by all research groups with vertically oriented flow cells. Because a detection volume of more than 60 μL leads to severe HPLC degradation for small analytical columns, a vertically oriented flow cell with a detector volume of 50–60 μL and utmost NMR sensitivity seems to be the best approach for stopped- and continuous-flow HPLC-NMR. An increase in sensitivity is possible with Helmholtz coils built with susceptibility-corrected wire directly fixed at the glass cell. This recently constructed probe exhibits excellent sensitivity and good resolution. The 2D correlated spectrum shown in Figure 9.27 was recorded with this cell design. The solvent signal problem can be partially solved by the use of a high-dynamic-range receiver incorporated in newer NMR instruments. To get rid of the non-linearity of transverse magnetization excited by binomial signal suppression techniques (Figure 9.20), the application of solvent presaturation by a special de-coupling coil seems to be favorable.

Further applications of on-line HPLC NMR coupling will be a combination of continuous- and stopped-flow techniques. First, a continuous-flow run will give information on the ^{1}H-NMR signals of the chromatographic peaks. Storing the retention data in an automation program, a series of interrupted HPLC separations will be performed, bringing each interesting peak in the NMR detection volume. Corresponding 2D spectra of all interesting peaks can be recorded. Thus, extremely valuable information about unknown compounds can be obtained overnight.

Another development will be the continuous-flow registration of nuclei such as ^{19}F,[66] ^{13}C,[67–69] and ^{31}P. Here, problems with solvent signal do not exist. Because continuous-flow inverse ^{1}H$\{^{13}$C$\}$ experiments proved to deliver valuable information,[70] the development of an inverse continuous-flow probe will considerably enhance the information content of on-line HPLC-NMR coupling.

9.9. References

1. R. P. W. Scott, *Liquid Chromatography Detectors,* 2nd edn., Elsevier: Amsterdam (1986).

2. H. C. Dorn, *Anal. Chem.* **56** (1984) 747A.

3. D. A. Laude, Jr., and C. L. Wilkins, *Trends Anal. Chem.* **5** (1986) 230.

4. K. Albert and E. Bayer, *Trends Anal. Chem.* **7** (1988) 288.

5. A. I. Zhernovoi and G. D. Latyshev, *NMR in a Flowing Liquid,* Consultants Bureau: New York (1965).

6. C. A. Fyfe, M. Cocivera, and S. W. Damji, *Acc. Chem. Res.* **11** (1978) 277.

8. P. M. Dennis, G. J. Bene, and R. C. Exterman, *Arch. Sci.* **5** (1952) 32.

9. A. L. Bloom and J. N. Shoolery, *Phys. Rev.* **90** (1953) 358.

10. C. A. Fyfe, M. Cocivera, and S. W. H. Damji, *J. Am. Chem. Soc.* **97** (1975) 5707.

11. M. Cocivera, C. A. Fyfe, A. Effio, S. P. Vaish, and H. E. Chen, *J. Am. Chem. Soc.* **98** (1976) 1573.

12. C. A. Fyfe, C. D. Malkiewich, S. W. H. Damji, and A. R. Norris, *J. Am. Chem. Soc.* **98** (1976) 6983.

13. M. Cocivera and A. Effio, *J. Am. Chem. Soc.* **98** (1976) 7371.

14. C. A. Fyfe, M. Cocivera, S. W. H. Damji, T. A. Hostetter, D. Sproat, and J. O'Brien, *J. Magn. Reson.* **23** (1976) 377.

15. C. A. Fyfe and L. Van Veen, *J. Am. Chem. Soc.* **99** (1977) 3366.

16. C. A. Fyfe, S. W. Damji, A. Knoll, and F. A. Forte, *Can. J. Chem.* **55** (1977) 1468.

17. C. A. Fyfe, S. W. Damji, and A. Knoll, *J. Am. Chem. Soc.* **101** (1979) 9519.

18. C. A. Fyfe, S. W. Damji, and A. Knoll, *J. Am. Chem. Soc.* **101** (1979) 958.

19. J. E. Singer, *Science* **130** (1959) 1652.

20. O. C. Morse and J. R. Singer, *Science* **170** (1970) 440.

21. T. Grover and J. R. Singer, *J. Appl. Phys.* **42** (1971) 938.

22. J. R. Singer, *Science* **175** (1972) 794.

23. J. H. Battocletti, J. H. Linehan, S. J. Larden, A. Sances, R. L. Bowman, V. Kudravcev, W. R. Genthe, R. E. Halback, and S. M. Evans, *IEEE Trans. Bio-Med. Eng.* **19** (1972) 403.

24. D. W. Arnold and L. E. Burichart, *J. Appl. Phys.* **36** (1965) 870.

25. W. S. McCormick and W. P. Birkenmeirer, *Rev. Sci. Instrum.* **40** (1960) 346.

26. S. Forsen and J. Rupprecht, *J. Chem. Phys. E* **2** (1969) 292.

27. W. L. Rollwitz and G. A. Persyn, *J. Am. Oil Chem. Soc.* **48** (1971) 59.

28. R. J. Hayward, K. J. Packer, and D. Tomunson, *J. Mol. Phys.* **23** (1972) 1083.

29. J. Kumar and V. Kumar, Science **175** (1972) 794.

30. D. W. Jones and T. F. Child, *Advances in Magnetic Resonance,* J. S. Waugh, ed., Vol. 8, Academic: New York p. 138 (1976).

31. J. J. Grimaldi, J. Baldo, C. McMurray, and B. D. Sykes, *J. Am. Chem. Soc.* **94** (1972) 7461.

32. J. J. Grimaldi and B. D. Sykes, *Rev. Sci. Instrum.* **46** (1975) 1201.

33. J. J. Grimaldi and B. D. Sykes, *J. Am. Chem. Soc.* **97** (1975) 273.

34. N. Watanabe and E. Niki, *Proc. Jpn. Acad. Ser. B* **54** (1978) 194.

35. E. Bayer, K. Albert, M. Nieder, E. Grom, and T. Keller, *J. Chromatogr.* **186** (1979) 497.

36. J. F. Haw, T. E. Glass, D. W. Hausler, E. Motell, and H. C. Dorn, *Anal. Chem.* **52** (1980) 1135.

37. E. Bayer, K. Albert, M. Nieder, E. Grom, and A. Zhu, *Fresenius Z. Anal. Chem.* **304** (1980) 111.

38. J. Buddruss and H. Herzog, *Org. Magn. Reson.* **13** (1980) 153.

39. J. Buddrus, H. Herzog, and J. W. Cooper, *J. Magn. Reson.* **42** (1981) 453.

40. J. Buddrus and H. Herzog, *Org. Magn. Reson.* **15** (1981) 211.

41. J. F. Haw, T. E. Glass, and H. C. Dorn, *Anal. Chem.* **53** (1981) 2327.

42. J. F. Haw, T. E. Glass, and H. C. Dorn, *Anal. Chem.* **53** (1981) 2332.

43. E. Bayer, K. Albert, M. Nieder, E. Grom, G. Wolff, and M. Rindlisbacher, *Anal. Chem.* **54** (1982) 1747.

44. J. F. Haw, T. E. Glass, and H. C. Dorn, *J. Magn. Reson.* **49** (1982) 22.

45. J. F. Haw, T. E. Glass, and H. C. Dorn, *Anal. Chem.* **55** (1983) 22.

46. J. Buddrus and H. Herzog, *Anal. Chem.* **55** (1983) 1611.

47. D. A. Laude, Jr., and C. L. Wilkins, *Anal. Chem.* **56** (1984) 2471.

48. K. Albert, M. Nieder, E. Bayer, and M. Spraul, *J. Chromatogr.* **346** (1985) 17.

49. D. A. Laude, Jr., R. W.-K. Lee, and C. L. Wilkins, *Anal. Chem.* **57** (1985) 1281.

50. D. A. Laude, Jr., R. W.-K. Lee, and C. L. Wilkins, *Anal. Chem.* **57** (1985) 1464.

51. D. A. Laude, Jr., and C. L. Wilkins, *Anal. Chem.* **59** (1987) 546.

52. L. A. Allen, T. E. Glass, and H. C. Dorn, *Anal. Chem.* **60** (1988) 390.

53. K. Albert, M. Kunst, E. Bayer, M. Spraul, and W. Bermel, *J. Chromatogr.* **463** (1989) 355.

54. K. Albert, M. Kunst, E. Bayer, H. J. de Jong, P. Genissel, M. Spraul, and W. Bermel, *Anal. Chem.* **61** (1989) 775.

55. D. I. Hoult and R. E. Richards, *J. Magn. Reson.* **24** (1976) 71.

56. R. P. W. Scott and P. Kucera, *J. Chromatogr. Sci.* **17** (1985) 346.

57. M. Spraul, M. Hofmann, H. Glauner, J. Gans, and K. Albert, Braker Report **2** (1990) 12.

58. C. Bauer, R. Freeman, T. Frenkiel, J. Keeler, and A. Shaka, *J. Magn. Reson.* **58** (1984) 442.

59. G. M. Clore, B. J. Kimber, and A. M. Gronenborn, *J. Magn. Reson.* **54** (1983) 170.

60. D. L. Turner, *J. Magn. Reson.* **54** (1983) 146.

61. P. J. Hore, *J. Magn. Reson.* **55** (1983) 283.

62. P. J. Hore, *J. Magn. Reson.* **56** (1984) 535.

63. H. Engelhardt, *Hochdruckflüssigkeitschromatographie,* 2nd edn., Springer-Verlag; Berlin (1977).

64. T. E. Glass and H. C. Dorn, *J. Magn. Reson.* **51** (1983) 527.

65. T. E. Glass and H. C. Dorn, *J. Magn. Reson.* **52** (1983) 518.

66. L. A. Allen, M. P. Spratt, T. E. Glass, and H. C. Dorn, *Anal. Chem.* **60** (1988) 675.

67. E. Bayer and K. Albert, *J. Chromatogr.* **312** (1984) 91.

68. K. Albert, G. Kruppa, K.-P. Zeller, E. Bayer, and F. Hartmann, *Z. Naturforsch.* **39c** (1984) 859.

69. K. Albert, E. Dreher, H. Straub, and A. Rieker, *Magn. Reson. Chem.* **25** (1987) 919.

70. K. Albert, J. L. Sudmeier, M. Anwer, and W. W. Bachovchin, *Magn. Reson. Med.* **11** (1989) 309.

Index

Gas chromatography/Fourier transform infrared
   spectrometry, 128
Glassy carbon electrodes, 96–97, 105, 106
Gold electrodes, 97, 101, 106
Gradient elution, 18
Green's function, 112

Half-wave potential, 93
Herbicide and herbicide metabolite applications
   of LC/MS, 177–79
High-performance micro separations, laser-based
   detection in, See Laser-based detection in
   high-performance micro separations
Homologous polynucleotides, detection of, 44
Hydrocarbon application of HPLC-NMR on-line
   coupling, 216–17
Hydrodynamic voltammograms (HDVs), 93–94
Hydrophilic binding sites, 86
Hydrophobic binding sites, 86

Immobilized phosphorescence detection, 29–34
Immunoaffinity columns, 67
Indirect detection strategies
   laser-based, 8–9, 19–22
   phosphorescence, See Room-temperature
   phosphorescence (RTPL) detection
Indocyanine green, 83–84
Inelastic scattering, 17
Infrared spectrometry, See Fourier transform in-
   frared spectrometry/HPLC
Instrument response function, 1–4
Interferences, 53
Intracavity applications of laser diodes, 82–83
Intracavity photothermal HPLC detectors, 123
Inverse continuous-flow nuclear magnetic reso-
   nance probes, 227
Iodoacetamidesalicylic acid, 47–48
Ion spray ionization LC/MS interface, 173,
   175–76
Iron magnet systems, 204

Labeled antibodies, 71
Lanthanide luminescence detection, 28, 39–53
   potential of, 41–42
   quenched, 50–53
   sensitized, 42–50
   spectroscopy of lanthanides, 39–41
Laser-based detection in high-performance micro
   separations, 1–23
   analytical figures of merit, 4–7
   approaches, 11–17
   concentration and mass sensitivity, 5–7
   direct and indirect, 8–9
   future developments, 23

instrument response function, 1–4
optical imaging and detection strategies, 7–8
photothermal detectors, See Photothermal de-
   tectors
review of trends, 10
universal detection approaches, 17–23
varieties of micro separations, 9–10
Laser diodes, 80–83
Laser-induced fluorescence (LIF), 8–10, 13–14,
   21; See also Near infrared HPLC
LC/MS techniques, See Mass spectrome-
   try/liquid chromatography techniques
Light-emitting diodes (LEDs), 80–81
Light scattering detector (LALS), 10
Limiting current region, 93
Limit of detection, 3–4, 72
Linear dynamic range, 3
Lipid applications of LC/MS, 187, 188
Liquid chromatography with electrochemistry
   (LCEC), See Electrochemical detection
Lockin amplification, 7
Long-lived luminescence detection, 27–53. See
   also Lanthanide luminescence detection;
   Room-temperature phosphorescence (RTPL)
   detection
Luciferase, 60–62
Luciferin, 60, 61
Lucigenin chemiluminescence, 69–70
Luminescence detection, See Chemiluminescent
   detection; Long-lived luminescence detec-
   tion
Luminescence lifetime, 39–40
Luminol chemiluminescence, 65–69

Magnet field homogeneity, 213
Magnet homogeneity shimming, 213
Magnetization curves, 199–200
Maleimidylsalicylic acid, 47–48
Mass sensitivity, 5–7
Mass spectrometry/liquid chromatography techniques, 163–90
   continuous-flow fast atom bombardment interface, 171–74, 177, 178, 180–85, 187–88,
   190
   direct liquid introduction interface, 165–68,
   177, 178, 182, 185
   drug and drug metabolite applications, 176–
   77
   electrospray ionization interface, 173, 175–
   76, 177, 180–85, 190
   environmental sample applications, 181
   herbicide and herbicide metabolite applications, 177–79
   ion spray ionization interface, 173, 175–76